Corrosion of archaeological and heritage artefacts

This special issue of *Corrosion Engineering Science and Technology* is dedicated to the study of corrosion of objects from historical sites. The issue contains contributions from the 2009 EUROCORR session on 'Corrosion of Archaeological and Heritage Artefacts' organised by the European Federation of Corrosion's working party and commissioned articles on other key issues. The objective is to give the reader a broad understanding of corrosion of ancient materials, for the most part metal but also glass.

In publishing different kinds of articles we hope to shed light on a range of analytical approaches related to the study of the complex systems that make up historical artifacts. In order to arrive at an understanding of the nanometric organisation of rust layers and interphases, such studies must be approached on a macroscopic scale. Techniques used include; macrophotography, synchrotron radiation and transmission electron microscopy (TEM) that ensure results that are both exhaustive and representative of particular observations. This issue proposes to demonstrate the wealth of approaches possible in the study of the corrosion of ancient materials.

In addition, understanding of century-old corrosion systems and the development of effective conservation treatments clearly require an advanced scientific approach based on the informing of mechanisms at various scales. This calls for dialog at the European level between teams able to produce innovative scientific studies of the objects under consideration; a task all the more challenging because such artefacts are composed of materials that are often more heterogeneous than in contemporary constructions and less frequently found. The combination of these factors adds to the difficulty of alteration studies and requires a collaborative European and even a global effort in determining the scientific priorities to be addressed over a given period of time. Publications such as this issue make it possible to identify crucial points in the field.

It must be noted that century-old materials pose both all the problems associated with corrosion of modern day materials as well as specific issues related to preserving materials (insofar as possible) during conservation treatment. Considerations such as these, amount to what could practically be considered a materials field in itself. This is precisely what was demonstrated during forums held on this theme at the 2009 EUROCORR conference. Moreover, ancient materials research is eminently interdisciplinary involving archaeological sites, museums, laboratories and even synchrotron facilities. Here again, coordination and international cooperation is key and the ECF's Working Party 21 is committed to achieving such a goal in the years to come, a goal which can only be accomplished through teamwork amongst institutions striving to meet these objectives. It is only through such concerted efforts that it will be possible to 'fill the gap' between restoration workshops and scientific laboratories.

I would like to conclude this editorial with a dedication to Professor R. Balasubramaniam, a colleague and a friend who passed away unexpectedly last year. A driving force at IIT Kanpur, he carried out research on the corrosion of historical artefacts with both passion and immense scientific competency. He was a constant supporter of both global research and international scientific coordination. He was on the editorial board of this journal and it is thanks to his efforts, in part, that this special edition was able to appear.

Philippe Dillmann, *Guest editor*
Chairman of WP21 of European Federation of Corrosion
UMR5060 and UMR9956 CNRS and CEA - France

DOI 10.1179/147842210X12855853971160

Intentional patina of metal archaeological artefacts: non-destructive investigation of Egyptian and Roman museum treasures

M. Aucouturier[1], F. Mathis*[2], D. Robcis[1], J. Castaing[1], J. Salomon[1], L. Pichon[1], E. Delange[3] and S. Descamps[4]

This paper describes microstructural analyses by X-ray portable diffraction and microdiffraction on intentional patina of the bronze museum objects from antique Egypt and the Roman Empire. They bring evidence of the presence in the true black bronze patinas of metallic gold and/or silver presumably as nanoparticles. Three other Egyptian patinas not belonging to black bronze are characterised. Apparent black patination on a Roman scalpel handle is discussed. The discovery of a new patination procedure on Roman artefacts from the Louvre museum is also related, based on intentional high temperature oxidation to obtain a dark patina on a lead bronze object. A presence of lead carbonate cerussite is an important observation.

Keywords: Egyptian black bronze, Roman Corinthian bronze, Thermal patina, Ion beam analyses, X-ray microdiffraction

This paper is part of a special issue on corrosion of archaeological and heritage artefacts

Introduction

Research into the original appearance of ancient metal artefacts[1–6] presents major difficulties owing to corrosion or restoration.

A specific black intentional patination was recently identified[3,4] in a number of Egyptian and Roman artefacts and compared with the chemical patination of Japanese *shakudo* Cu–Au alloys. This patina is made of Cu_2O oxide (cuprite), which takes its black colour from the presence of Au in the alloy, called 'black bronze', *Corinthian bronze* or *shakudo*. The oldest 'black bronze' known dates from the Egyptian Middle Kingdom.[7,8] A complete review may be found elsewhere.[5] The typological observations done by the first Egyptologists[9] were not sufficient to define precisely the typical black bronzes, and a rigorous definition was given through physicochemical studies,[5,10,11] which provided scientific criteria to identify definitely the intentionally patinated black bronzes and differentiate them from pieces neither intentionally patinated nor possibly patinated with another recipe:[1,10,12] the patina is a cuprite (Cu_2O) layer containing Au and/or Ag.

The present paper has two aims: one was to attempt a further characterisation of the black bronze intentional patina, using investigation methods complementary to the ion beam analyses reported in our previous works; the second one was to describe the evidence of intentional patination techniques not belonging to the 'black bronze' or 'Corinthian bronze' class.

Objects

All objects belong to the Louvre museum.

Six artefacts are from the Egyptian Antiquities Department (Table 1 and Fig. 1) selected among 16 previously reported.[10] Among them, three are definitely recognised as black bronzes: the Henouttauy sistrum (E 11201), the Siamon Sphinx (E 3914) and the double Aegis (N 4302). The Harpocrates figurine E 7735 has a part made in black bronze, its lock of hair. Three seem to be intentionally patinated but do not belong to the 'black bronze' class: the Harpocartes figurine body, inlays on the Amon statue body (N 3547) and the feminine statuette (E 27430). The Amon statue base has a hieroglyph inlaid with black bronze, but the statue itself is inlayed with orange red inlays.

From the Greek, Etruscan and Roman Antiquities Department are three artefacts dated from the Roman Empire (Table 2 and Fig. 2).

First, the Vaison-la-Romaine inkpot (Bj 1950) is described in several publications.[1,5,11] It is decorated with various inlays, particularly with inlays in Corinthian bronze: the *himation* (coat) of Venus and Adonis figures and the wings of the Eros figures.

Second, a scalpel handle (Br 2416) in Cu–6Zn–3Pb alloy (wt-%) has black patinated inlays; the black inlays were themselves originally inlayed with presently missing Ag inlays.

Third, a bronze herm from a balustrade with a satyr head and a fragment of another similar pillar (Br 4648-1

[1]C2RMF (CNRS UMR171), Palais du Louvre, Porte des Lions, 15 quai François Mitterrand, 75001 Paris, France
[2]Centre Européen d'archéométrie, Université de Liège, 15 Allée du Six Août, Sat Tilman B 15, 4000 Liège, Belgium
[3]Musée du Louvre, DAE, Palais du Louvre, 75001 Paris, France
[4]Musée du Louvre, DAGER, Palais du Louvre, 75001 Paris, France

*Corresponding author, email Marc.aucouturier@culture.gouv.fr

Published by Maney on behalf of the Institute
Received 9 December 2009; accepted 8 March 2010
DOI 10.1179/147842210X12710800383567

a harpocrates; *b* Sphinx of Siamun; *c* Henuttaoui sistrum; *d* double Aegis; *e* feminine statuette; *f* Amun statue

1 Objects from Egyptian Antiquities Department of Louvre museum (see Table 1) (C2RMF, D. Bagault; scale bar=2 cm)

and 4648-2) are intentionally patinated in black and present different Cu and Ag inlays. Both the pillar body and the head are patinated. They may be linked to series of objects (*Herms*) from the Roman Empire, often in marble described for instance in Ref. 13, with no mention of any possible intentional patina.

Table 1 Description of six investigated Egyptian artefacts: ▲ are black bronzes (Hmty-km)

Objects	Inventory no.	Period	Patina	Description
Harpocrates body, hair lock (▲)	E 7735	Late eighteenth–early nineteenth dynasty (1330–1210 BC)	Black patina on lock of hair; dark green on the body	Harpocrates (child king) seated, naked, without arms; lost inlaid eyes. The body has a dark patina, with plated Ag nails. Black hair lock inlaid with Au strips
Sphinx of Siamun (▲)	E 3914	Twenty-first dynasty (978–959 BC)	Black background	Sphinx inscribed for King Siamun with hands supporting a table of offerings. Black background with a Au inlaid decoration
Double Aegis (▲)	N 4302	Twenty-first dynasty (1069–945 BC)	Black background	Tiny double Aegis assembled with its collar counterpoise inlaid with three kinds of Au
Henuttaui sistrum* (▲)	E 11201	Third Intermediate Period (1069 to ~800 BC)	Black patina on the entire handle and on the urei	Sistrum inscribed with the name of the musician Henuttaui. Cylindrical handle bearing a double head of Hathor surmounted by an ureus frieze and framed by two added urei
Feminine statuette	E 27430	Third Intermediate Period?	Black background	Feminine statuette, 12·7 cm high, without head, arms and feet, decorated with Au inlays featuring a bead netting and a collar
Statue of Amun base inlay (▲)	N 3547	Late Period (664–332 BC)	Black hieroglyph; orange patina on beard collar, belt and *ousekh* collar	Amun standing, left arm extended, with loincloth, Amun crown, false beard and *ousekh* collar. Base inlaid hieroglyphic inscription with black patina

a Vaison-la-Romaine inkpot; *b* scalpel handle; *c* herm (partial view) and fragment

2 Objects from Department of Greek, Etruscan and Roman Antiquities, Louvre (see Table 2): scale bar=1 cm

Instrumentation and methods of analysis

As related in previous publications, the bulk metals were analysed either by inductively coupled plasma atomic emission spectroscopy or by particle induced X-ray emission (PIXE) on chips sampled by microdrilling.[1,5,10,11] The metal of the balustrade fragment was analysed by PIXE directly on cleaned areas; the given composition is thus approximate because of possible distortion by corrosion phenomena.

The inlays and intentionally patinated areas had to be analysed non-destructively; this was done directly on the surface by PIXE under 3 MeV protons or under 6 MeV alpha particles. The thickness of the patina was measured by Rutherford backscattering spectrometry (RBS). All details about the ion beam analyses conditions on the particle accelerator AGLAE of C2RMF may be found elsewhere.[1,5,10,11,14] It is important to emphasise that the PIXE analysis allows quite low detection limits for elemental metal analysis, e.g. <100 wt ppm for the Ag and Au contents in Cu alloys and in the oxide patinas.

The compositions of Egyptian artefacts are discussed from the Egyptology viewpoint in the catalogue of the exhibition held in New York in October 2007 to February 2008.[15]

The detection limit of PIXE for light elements (O, N, Cl, Na, Al, etc.) is not good, and this is one justification of the use of X-ray diffraction to identify oxides, corrosion products, patina components and similar compounds.

Two newly developed totally non-destructive investigation methods were used here.

One is a portable X-ray diffraction equipment,[16] designed to perform *in situ* X-ray diffraction on museum artefacts. The X-ray source is a Cu K_α source equipped with a polycapillary semilens providing a 4 mm diameter beam.

The second equipment is a recently designed X-ray microdiffraction equipment.[17] The source is a Rigaku microfocus X-ray tube. A multicapillary system and a collimator provide an incident parallel monochromatic Cu K_α X-ray beam with a diameter of 200 μm and a very small divergence. The X-ray flux is up to 2×10^8 photons/second, and the arrangement allows displaying

Table 2 Description of three investigated Roman Empire artefacts

Objects	Inventory no.	Period	Patina	Description
Vaison-la-Romaine inkpot	Bj 1950	First century AD	Corinthian bronze inlays: Venus and Adonis' coats, Eros wings	Small brass inkpot inlayed with Ag, Cu, Au and Corinthian bronze figures portraying a mythological scene with Venus, Adonis, Psyche and 6 Eros
Scalpel handle	Br 2516	First to second centuries AD	Black intentional patina on inlays	Quaternary Cu alloy piece with black inlays, themselves inlayed probably with Ag
Balustrade pillar, pillar fragment	Br 4648	Roman Empire	Black intentional patina on the pillar body and satyr head	Bronze herm with a satyr head on its top, inlayed with Cu foliated decoration, Cu satyr lips, Ag satyr eyes and floral decoration, on a black background

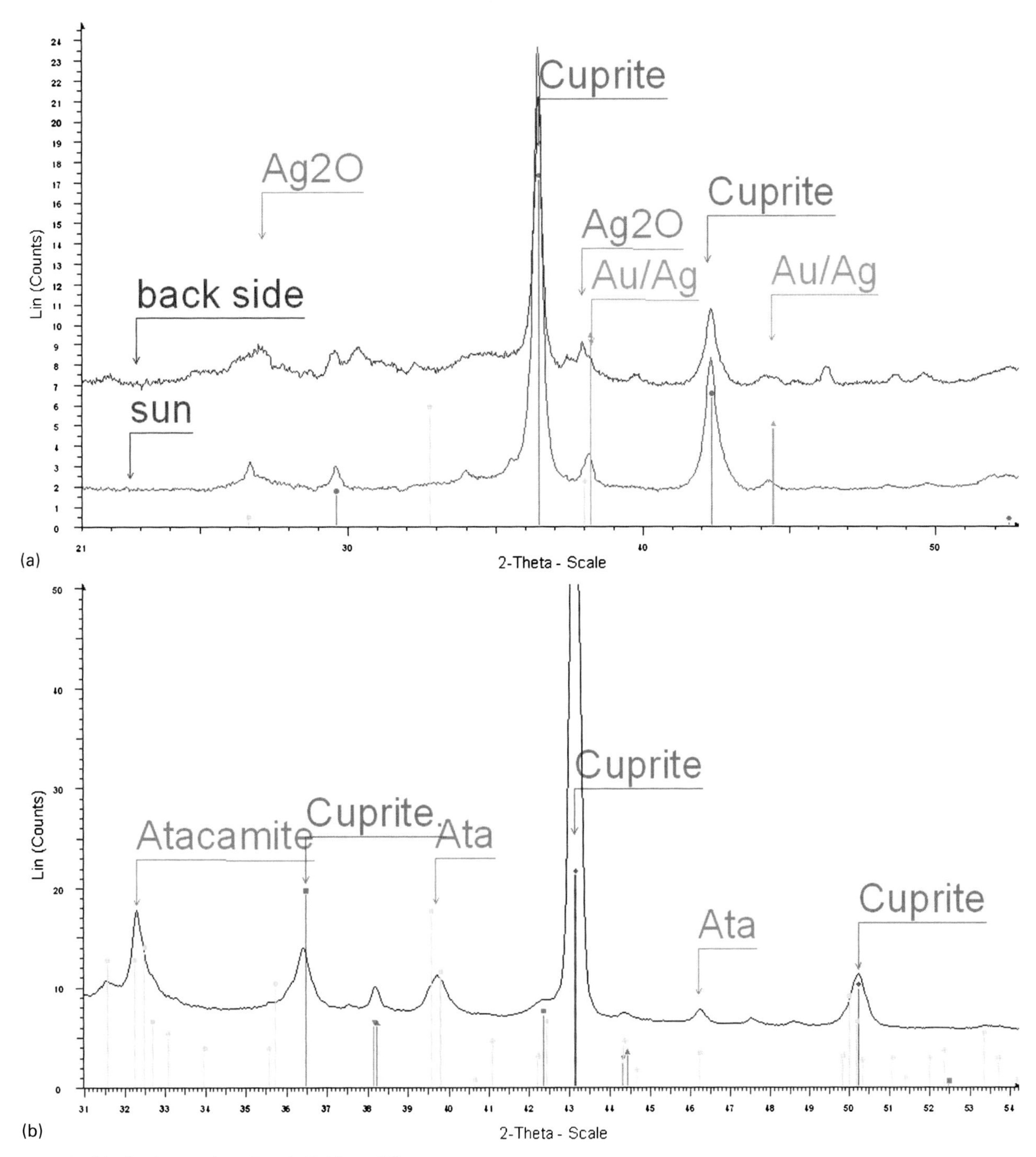

a double Aegis counterpoise; *b* Sphinx of Siamun

3 X-ray microdiffraction patterns on black patina of two objects

in front of the beam any kind of object, including large museum artefacts.

With both equipments, the diffraction pattern is detected by an imaging two-dimensional plate, post-treated by the FIT2D code.[18] The exposure times to obtain the diagrams shown in this paper are currently <30 min.

As for any X-ray diffraction analysis, the detection limit of minor compounds is of the order of a few volume per cent in the analysed volume. Here, the incidence angle of the X-ray beam can be set smaller than 10°, thus favouring the detection of surface compounds; the analysed depth is of the order of 10 μm.

Results on 'black bronzes': interpretation

The analysis of the base metal of black inlays or very small parts (Harpocrate's lock of hair, statue of Amun base inlay and Vaison-la-Romaine 'Corinthian

Table 3 Bulk metal composition in weight per cent of black patinated items: PIXE on drilled microsamples[1,10]

Object	Cu	Sn	Zn	Pb	As	Au	Ag	Fe
Siamun Sphinx	93·8	0·9	nd	0·77	1	1·3	1·5	0·66
Double Aegis	86	6	nd	0·3	2	0·8	3	1·3

Table 4 Composition of black patinas in weight per cent (average content of main elements)[1,5,10]

Objects	Cu	Sn	Zn	Pb	As	Fe	Au	Ag	Cl	S
Harpocrates lock of hair*	79	7	nd	4	2	nd	6·7	nd	2·9	2·6
Sphinx of Siamun†	90	2·5	0·4	1	1·5	0·5	0·5	3	3	2
Double Aegis counterpoise†	40–70	10–19	nd	4–12	4	5	2–3	2–4	2–5	1
Sistrum of Henuttauy handle†	85	7	0·3	0·2	2·5	0·1	2·5	1·5	2	0·5
Sistrum of Henuttauy urei†	90	3	0·2	0·5	0·8	0·5	1·5	0·5	4	1
Amun hieroglyph base*	92	3	nd	3	0·4	0·3	0·02	0·05	Yes	Yes
Inkpot of Vaison-la-Romaine 'Corinthian bronze' inlays	90	nd	nd	0·9	3	nd	1	3	1	1

*Surface PIXE analysis under 3 MeV protons.
†Surface PIXE analysis under 6 MeV alpha particles.

bronze' inlays) could not be performed because it was not possible to sample that metal. Because of its exceptional good state of conservation, that analysis was also not possible for the bulk metals of the various patinated parts of the Henuttaui sistrum (handle and added urei). Table 3 recalls the results (averaged values for the principal elements) obtained on microsamples of black patinated items, when sampling was allowed.[1,10] For the other black patinated items, only a qualitative appreciation of the nature of the bulk metal could be inferred from the composition of the patina (Table 4).

The analyses on black patinas are given in Table 4. They were obtained by direct PIXE analysis on the surface, either under 3 MeV protons or under 6 MeV alpha particles. The latter configuration allows a PIXE analysis of the only patina, without any influence of the underlying bulk metal.[1,19] The thicknesses of the patinas, measured by RBS are >10 μm except for the Sphinx of Siamun (4·5 μm) and the Amun base hieroglyph (3 μm).

Examples of diffractograms on black patinas are given in Fig. 3. The patina is mainly constituted of cuprite Cu_2O as previously reported.[1,10,11] As a new result here, diffraction peaks of metallic Ag or Au are evidenced on almost all patterns. The lattice parameters of metallic Au and Ag (0·4079 and 0·4086 respectively) are very similar, and it was not possible to differentiate them. Sometimes, the presence of Ag oxide could also be possible.

The patina of 'black bronze', only constituted of naturally red cuprite Cu_2O, becomes black when the metal contains Au or/and Ag. Some authors suggested that this could be due to the presence of Au nanoparticles and the known optical absorbing effect of such nanoparticles, due to surface plasmon phenomena.[20] The present observation brings a positive argument for that interpretation.

Table 5 Analysis by PIXE on clean area of scalpel handle black inlays[1,5] (approximate weight per cent values)

	Cu	Sn	Pb	As	Fe	Ag
Scalpel black inlay	98	0·3	0·3	0·4	0·5	0·2

Results on scalpel handle: discussion

The Roman scalpel handle black inlays may be suspected to be in black bronze, called 'Corinthian bronze' at that period.

The alloy constituting the inlays could be locally analysed semiquantitavely by PIXE on areas where the patina is missing.[1,5] The results, to be considered as an approximate estimation of the alloy composition, are given in Table 5. The composition of the patina, also obtained by PIXE analysis, is given in Table 6.[1,5]

The values obtained for the black patina (Table 6) are separated into two groups: one (first line) corresponds to the metallic elements visibly belonging to the patina and summed to 100%; the second one (second line with a somewhat arbitrary quantitative level) corresponds to elements obviously coming from pollution by the burying environment of the object. That presentation explains why the total concentration sum is not 100%.

The alloy used does not contain any Au addition but a small Ag addition (0·2 wt-%). The Ag content is enriched in the patina (1 wt-%) as compared to the underlying alloy.

The microdiffraction pattern on a black inlay confirms cuprite Cu_2O as the main patina component. Other corrosion compounds are also present in smaller quantity, as malachite $Cu_2CO_3(OH)_2$. No metallic Ag is found. A possible, not certain, presence of Ag oxide may be supposed. If one refers to the definition proposed in previous publications,[1,3,5,10] this patina satisfies two of the criteria for being a 'black bronze' patina: it is constituted of cuprite and contains noticeable amount of noble metal, here Ag. However, it does not contain noble element in metallic state: Au is absent and Ag is probably oxidised.

One may be cautious before considering that these inlays are belonging to the family of the 'black bronzes', as the Ag content of the inlay alloy might be considered as a natural Cu impurity content, although it is higher than the Ag content of the alloy of the bulk object, not detectable, i.e. <100 wt ppm. However, the crystallographic nature of the patina, cuprous oxide, its black colour obviously intentional (instead of red, natural colour of cuprite Cu_2O) and the Ag enrichment of the patina as compared to the substrate are arguments in favour of a specific patination recipe of the 'black bronze' class.

Table 6 Analysis by PIXE analysis on black patina of scalpel inlays (average values in weight per cent)[1,5]

	Cu	Sn	Pb	As	Fe	Ag	Al	Si	P	S	Cl	K	Ca
Metallic elements	94	1·5	0·9	0·8	1·5	1	~1						
Pollution elements								30–60	~1	~1	~1	~0·5	~2

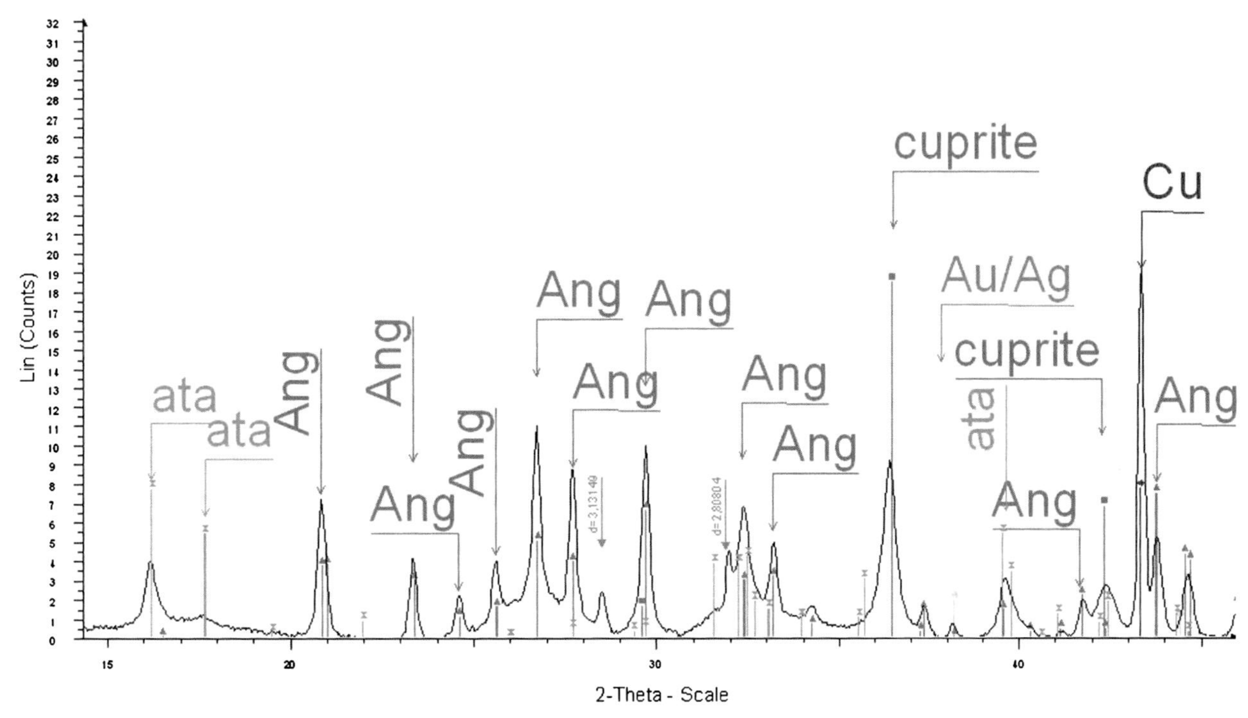

4 Microdiffraction pattern of dark patina of feminine statuette: ata, atacamite; Ang, anglesite

Results on other (not 'black bronze') Egyptian patinas: discussion

The statue of Amon is inlayed with pieces (eye pupils, beard collar, *ousekh* collar of pearls and belt) with a colour presently orange red, not very different from the neighbouring corrosion; the Harpocrates body is patinated in dark green colour; the feminine statuette is of dark colour in between the Au strings of the dress but contains neither Au nor Ag.

The compositions of the bulk metal of the Harpocrates and the feminine statuette are given in Table 7 (average composition by PIXE on chips of microdrilled samples).[1,10] The compositions of the patinas are given in Table 8.

The thicknesses of the patinas, measured by RBS, are all <10 µm.

The patina of the feminine statuette (Fig. 4) contains various Cu and lead compounds, as cuprite Cu_2O, atacamite $Cu_2Cl(OH)_3$ and anglesite $PbSO_4$. Careful optical microscope examination shows that it is indeed a mixture of corrosion products, which have 'pushed up' the neighbouring Au inlays to the surface. It is thus impossible to know the original aspect of the statuette surface, when it was fabricated.

The Harpocrates body is made of a very peculiar feature: a solid cast of a pure Cu–25 wt-%Au alloy, without any other addition. Its patina is a mixture of nantokite CuCl and atacamite $Cu_2Cl(OH)_3$. Au appears in the patina by PIXE analysis at 16 wt-% level instead of 25 wt-% for the bulk metal. Au is thus not enriched in the patina: the RBS spectrum shows that the Au content of the patina is much <1 wt-%. In fact, the patina is thin enough (<10 µm) to allow the proton beam to reach the underlying metal. Microdiffraction shows metallic Au peaks, but one could think that the diffraction peaks come in fact from a depth where the bulk alloy is depleted in Cu as a consequence of the surface chlorides formation. The absence of Au enrichment in the patina rules out the hypothesis of a 'black bronze'. Elaboration of chemical dark green chloride patina on Cu based alloys is described in known recipes, even in modern times.[21] It is impossible to know if the Harpocrates body patina is an authentic patina or if it has been applied in a recent period.

The inlays of the Amon statue (beard collar, *ousekh* collar and belt) are made with an alloy very different from the statue itself: the statue is in a common bronze (10 wt-%Sn and 0·3 wt-%Pb), whereas the inlays are in low alloyed Cu containing only 2 wt-%Sn, ~1 wt-%Pb and a rather high amount of bismuth (nearly 1 wt-%), which may be an interesting geochemical tracer.[22] The patina is mainly cuprite Cu_2O, explaining its colour. One may suppose that the inlays were originally patinated to appear different from the body, but it is impossible to precisely determine if their colour was originally black, as it could be supposed for instance for the eye pupils.

Table 7 Bulk metal composition in weight per cent of patinated Egyptian objects:[1,10] PIXE analysis on drilled microsamples

Objects	Cu	Sn	Zn	Pb	As	Au	Ag	Fe
Feminine statuette	87·5	5	nd	5	2	nd	0·2	nd
Harpocrates body	≈75	nd	nd	nd	nd	25	nd	nd
Amon statue	88	10	nd	0·3	0·2	nd	nd	0·2

Table 8 Patina composition of Egyptian items, weight per cent direct PIXE surface analysis under 3 MeV protons

Objects	Cu	Sn	Zn	Pb	As	Fe	Au	Ag	Cl	S	Bi
Feminine statuette dress	70	12	nd	15	3	0·6	nd	nd	10	4	nd
Harpocrates body	62	nd	nd	nd	nd	nd	16	nd	20	2	nd
Amun statuette inlays	95	2	nd	1·2	0·2	0·2	nd	nd	...	...	0·8

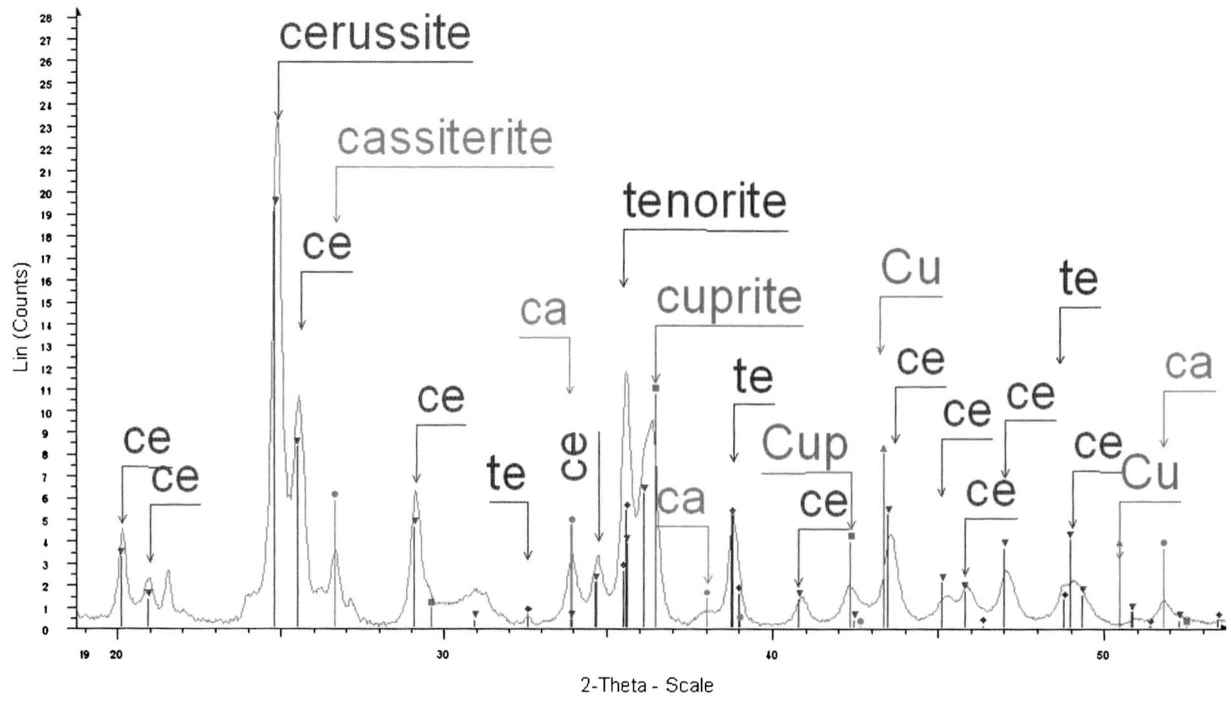

5 Microdiffraction pattern of black patina of balustrade fragment: ca, cassiterite; ce, cerussite; te, tenorite; cup, cuprite; Cu, Cu alloy

The role of the bismuth (intentional addition or Cu impurity) is not clear either. It is reminded (see above) that the base of that statue is inlayed (hieroglyphs) with patinated 'black bronze'.

Results on Roman bronze balustrade pillar: interpretation

The approximate semiquantitative composition of the balustrade part, obtained by PIXE analysis directly on an unpatinated area, is Cu–10Sn–10Pb–1Zn (wt-%).

Neither the base metal nor the patina contains Au or Ag (with a detection limit <100 wt ppm): this object is not made with Corinthian bronze. From the microdiffraction pattern (Fig. 5), the patina is mostly constituted of cuprite Cu_2O, tenorite CuO and cassiterite SnO_2. Black tenorite is the external component. A fundamental study of high temperature oxidation of Cu–Sn alloys[5,23] showed that the simultaneous occurrence of those three compounds can only be the consequence of a high temperature oxidation. Thanks to the RBS measurement of the respective thickness of these compounds, one deduces that the objects were heated under oxidising atmosphere at a temperature around 600°C. The same diagram was found on the pillar body and on the head's face, meaning that high temperature heating was not accidental.

Figure 5 reveals another important feature: the surface layer contains high amount of lead carbonate cerussite $PbCO_3$. The role of cerussite (of white natural colour) on the properties of the black thermal patina is yet not explained.

The bronze balustrade has been intentionally patinated in black by thermal oxidation at 600°C or higher. That thermal oxidation cannot be a consequence of an accidental fire or sacrificial event for the following reasons:

(i) the face of the satyr head is patinated exactly with the same layer as the body of the pillar and is in a perfect conservation state
(ii) red Cu inlays (e.g. the lips of the satyr head and a plant inlay stem on the pillar body) and white Ag inlays (e.g. the eyes white) have been inserted after the formation of the black patina
(iii) both analysed objects (the balustrade pillar fragment and the complete herm shown in Fig. 2*c*) do show exactly the same patina with the same composition.

Further experimentation is scheduled in order to try to understand the role of lead and the condition of formation of the lead carbonate.

Conclusions

This analytical study of various museum bronze artefacts exhibiting intentional surface patination shows that combination of several non-destructive investigation methods, including X-ray microdiffraction, leads to a very rich characterisation of ancient surface processing of cultural heritage artefacts by chemical or thermochemical means.

The microstructures of Egyptian and Roman black bronzes have been precisely described showing definitely the presence of metallic Au and/or Ag inside the patina, presumably as nanoparticles already suggested without experimental evidence by previous authors. This confirms the role of metallic particles on their optical properties.

The case of the black patinated inlays of the Roman scalpel handle is more ambiguous: the patina is, like for black bronze, constituted of cuprite, but it contains only a small amount of Ag in an oxidised state. The role of Ag on the patina colour is not yet elucidated.

Various new intentional ancient patination treatments were evidenced and characterised.

1. One is an orange red patina containing only cuprite Cu_2O on inlays of an Egyptian statue of Amon, applied on a particular low alloyed Cu containing bismuth as an important impurity.
2. Another one is a black patina obtained on a Roman bronze balustrade fragment by thermal treatment at a temperature around 600°C to produce tenorite.

The present investigation leads also to suggest that the black patina of the Egyptian female statuette was so much altered by corrosion that it is impossible to know what was the original aspect. It also brought a serious doubt about the original, if any, character of the patina of the Egyptian Harpocrates.

Acknowledgements

This paper is dedicated to the memory of Joseph Salomon, former chief of the C2RMF AGLAE accelerator team, who passed away in January 2009.

The conception and development of the X-ray microdiffraction device is due to Philippe Walter from C2RMF. Photographs are done by Dominique Bagault from C2RMF. Alessandra Giumlia-Mair kindly provided the reference[13] concerning bronze herms comparable to the Louvre balustrade pillar.

References

1. F. Mathis, J. Salomon, S. Pagès-Camagna, M. Dubus, D. Robcis, M. Aucouturier, S. Descamps and E. Delange: 'Corrosion patina or intentional patina?: contribution of non-destructive analyses to the surface study of copper-based archaeological objects', in 'Corrosion of metallic heritage artefacts: investigation, conservation and prediction of long term behaviour' (European Federation of Corrosion 48), (ed. P. Dillmann, *et al.*), 219–238; 2007, Cambridge, Woodhead Publishing.
2. A. Giumlia-Mair and S. Quirke: *Rev. Egyptol.*, 1997, **48**, 95–108.
3. P. Craddock and A. Giumlia-Mair: 'Hsmn-Km: Corinthian bronze, shakudo: black-patinated bronze in the ancient world', in 'Metal plating and patination', (ed. S. La Niece and P. Craddock), 101–127; 1993, London, Butterworth-Heinemann Ltd.
4. A. Giumlia-Mair and P. Craddock: 'Corinthium aes, das schwarze Gold der Alchimisten', *Antike Welt.*, 1993, **24**, 2–62.
5. F. Mathis: 'Croissance et propriétés des couches d'oxydation et des patines à la surface d'alliages cuivreux d'intérêt archéologique ou artistique', PhD thesis, Université Paris Sud XI, Orsay, Paris, France, 2005, available at: http://tel.ccsd.cnrs.fr/tel-00011255
6. S. La Niece, F. Sherman, J. Taylor and A. Simpson: *Stud. Conserv.*, 2002, **47**, 95–108.
7. A. Giumlia-Mair: 'Das Krokodil und Amenemhat III. aus el-Faiyum', *Antike Welt.*, 1996, **27**, 313–321.
8. A. Giumlia-Mair: 'Das Sichelschwert von Bälata-Sichem', *Antike Welt.*, 1996, **27**, 337.
9. J. D. Cooney: *Z. Ägypt. Sprache Alt.*, 1966, **93**, 43–47.
10. F. Mathis, E. Delange, D. Robcis and M. Aucouturier: 'HMTY-KM (black copper) and the Egyptian bronzes' collection of the Musée du Louvre', *J. Cult. Herit.*, 2009, **10**, 63–72.
11. S. Descamp-Lequime: 'L'encrier de Vaison-la-Romaine et la patine volontaire des bronzes antiques', *Monuments Piot*, 2005, **84**, 5–30.
12. F. Mathis, S. Descamps, D. Robcis and M. Aucouturier: 'An original surface treatment of copper alloy in ancient Roman Empire: chemical patination on a Roman strigil', *Surf. Eng.*, 2005, **21**, 346–351.
13. H. Wrede: 'Die spätantike Hermengalerie von Welschbillig, Untersuchung zur Kunsttradition im 4. Jahrhundert n. Chr. und zur allgemeinen Bdeutung des antike, Hermenmals'; 1972, Berlin, Verlag Walter de Gruyter.
14. T. Calligaro, J.-C. Dran and J. Salomon: 'Ion beam analysis', in 'Non-destructive microanalysis of cultural heritage materials', (ed. K. Janssens and R. V. Grieken), 227–276; 2004, Amsterdam, Elsevier.
15. E. Delange: 'The complexity of alloys: new discoveries about certain bronzes in the Louvre', in 'Gifts for the gods, images from the Egyptian temples', (ed. M. Hill), 39–49; 2008, New York, The Metropolitan Museum of Art, Yale University Press.
16. A. Gianoncelli, J. Castaing, L. Ortega, E. Doorhyee, J. Salomon, P. Walter, J.-L. Hodeau and P. Bordet: 'A portable instrument for *in situ* determination of the chemical and phase composition of cultural heritage objects', *X-Ray Spectrom.*, 2008, **37**, 418–423.
17. L. de Viguerie, L. Beck, J. Salomon, L. Pichon and P. Walter: 'Composition of Renaissance paint layers: simultaneous particle induced X-ray emission and backscattering spectrometry', *Anal. Chem.*, 2009, **81**, 7960–7966.
18. ESRF, available at: http://www.esrf.eu/computing/scientific/FIT2D/ (accessed on 18 November 2004).
19. F. Mathis, B. Moignard, L. Pichon, O. Dubreuil, J. Salomon: 'Coupled PIXE and RBS using a 6 MeV He-4 (2+) external beam: a new experimental device for particle detection and dose monitoring', *Nuclear Instrument and Method in Physics Research, section B*, 2005, **240**, 532–538.
20. R. Murakami: 'Japanese traditional alloys', in 'Metal plating and patination', (ed. S. La Niece and P. Craddock), 85–94; 1993, London, Butterworth-Heinemann Ltd.
21. R. Hughes and M. Rowe: 'Colouring, bronzing and patination of metals'; 1991, London, Thames and Hudson.
22. F. Cesbron, P. Lebrun, J.-M. Le Cléac'h and J. Deville: 'Minéraux du cuivre', Minéraux et fossiles, hors série no. 27, November 2008.
23. F. Mathis, M. Aucouturier and P. Trocellier: 'Explanation of tin role in the high temperature oxidation resistance of bronzes', in 'Copper', (ed. J.-M. Welter); 2006, Weinheim, Wiley VCH.

Electrochemically synthesised bronze patina: characterisation and application to the cultural heritage

J. Muller*, G. Lorang, E. Leroy, B. Laik and I. Guillot

The aim of the present study is to determine the role of tin on the corrosion behaviour of bronzes. For this purpose, different nuances of single phase α-Cu(Sn) are immersed in a 10^{-2} mol L^{-1} deaerated sulphate solution, buffered at pH 6·8. Then, the chemical–physical and electrochemical characterisation of an artificial patina, thin layer of corrosion products electrochemically formed at the surface of a 7 wt-% tin bronze are performed. First of all, the synthesised species are cathodically reduced. The successive steps of the potential–time curves, characteristic of the electrochemical reactions occurring at the electrode surface, are successfully calibrated with reference oxides. In order to complete the previous results, X-ray photoelectron spectroscopy (XPS) depth profiles are carried out by abrading progressively the oxidised surface. The deconvoluted XPS and Auger electron spectroscopy (AES) spectra speak in favour of an 'alloy/SnO_2/Cu_2O' layered structure similar to that reported in the literature. Analyses carried out on the oxidised samples by transmission electron microscopy (TEM) confirm this pattern.

Keywords: Electrochemistry, Bronze patina, Tin oxide, XPS–AES, TEM

This paper is part of a special issue on corrosion of archaeological and heritage artefacts

Introduction

In many studies performed on soil buried archaeological bronzes (Cu–Sn alloys),[1–3] or on bronzes exposed to the atmosphere,[4,5] copper containing compounds have been reported to be the main constituents of natural patinas. Thus, the corrosion mechanism of bronze has been usually assimilated to that of copper[6–8] even if the role of other alloying elements has been increasingly considered.[9,10]

More recently, a layer enriched in tin is observed on the bronze patina by characterisation techniques allowing the detection of less crystallised oxides. A corrosion mechanism based on copper selective dissolution was proposed by Robbiola *et al.*[11] This mechanism points out two types of bronze corrosion models. Type I structure can be regarded as resulting from generalised corrosion processes in a low aggressive medium associated with the formation of passive barrier-like layers enriched in tin which reduce dissolution.

The type II structure results from localised corrosion phenomena but also from generalised attacks due to a high dissolution rate. The main characteristic is the presence of high chloride amounts at the internal layer/alloy interface.

These models are based on archaeological artefacts investigations.[12] For the last 10 years, many studies have been performed to find the presence of a layer enriched in tin on samples artificially corroded, particularly in the chloride media.[13,14] A blocking adherent layer involving a bronze behaviour in the sulphate media different from that of copper was evidenced.[15] However, the nature of the tin compounds in this environment is for the moment still in discussion.

In this context, this study focuses on the methodology adopted, first, to obtain the type I corrosion structure in a 0·01 mol L^{-1} sulphate aqueous solution, and second, to analyse the different layers. First of all, effect of tin on the corrosion behaviour of bronzes is studied by immersing different nuances of single phase α-Cu(Sn). Then, the tin enriched layer on a synthetic α phase $CuSn_7$ alloy is investigated, by simulating corrosion and aging through anodic polarisation of the alloy, in a potential range close to the open circuit potential. The composition and the structure of the corrosion layers are studied by electrochemical methods, X-ray photoelectron spectroscopy (XPS) and transmission electron microscopy.

Experimental procedures

Specimen's elaboration

Bronzes are elaborated by melting Cu (copper rod, 12·7 mm, Goodfellow, 99·95%) and Sn (tin rod, 6 mm, Puratronic, 99·9985%; Alfa Aesar, Alfa Aesar, Ward Hill, MA, USA) in a water cooled carbon graphite crucible by electromagnetic high frequency (Celes generator) under

CNRS, University Paris XII, Paris, France

*Corresponding author, email muller@icmpe.cnrs.fr

Published by Maney on behalf of the Institute
Received 15 December 2009; accepted 17 March 2010
DOI 10.1179/147842210X12692706339265

helium atmosphere. A second induction melting in electromagnetic levitation has been performed, with an overheating more than 100°C above the liquidus temperature in order to ensure a perfect chemical homogeneity of the melt. The three different alloys containing 7, 11 and 14 wt-% tin are then homogenised under helium atmosphere at 750°C during 24 h. An annealing at 400°C during 2 h followed by successive alloys rolling until 60% of deformation and recrystallisation annealing at 500°C during 1 h are carried out.

A single phase α-Cu(Sn) solid solution, without segregations and with an annealed microstructure characterised by thermal twins and 10–30 μm sized grains, is obtained.

Flat samples of 0·5 mm thickness used for surface characterisation and 4 mm diameter rotating disc working electrodes are made from these alloys, as well as from pure copper and tin. They are previously embedded in an inert resin, mechanically polished up to 4000 SiC grade, rinsed with deionised water and air dried before immersion into the electrolyte.

Electrochemical experiments

All the electrochemical experiments are performed using an Autolab PGSTAT 30 (Eco Chemie, Utrecht, The Netherlands) potentiostat/galvanostat in a classical three-electrode cell (capacity: ~20 mL), with platinum as counterelectrode and a saturated calomel electrode (SCE) as reference electrode. The 10^{-2} mol L^{-1} sulphate electrolyte is prepared from H_2SO_4 (SDS, purity=95%) and 1,4-piperazinediethanesulfonic acid buffer reagent (disodium salt-PIPES, Aldrich, St Louis, MO, USA; pH-buffering agent with pKa=6·8) used in *ad hoc* proportions to buffer the solution at pH 6·8.

The cell temperature is regulated at a constant temperature with a thermostat (25·0±0·1°C). The electrodes are cathodically polarised for 50 s before any electrochemical treatments in order to remove the native oxide film using a pulsed method [multiple potential steps between −1·0 and 0·6 V(SCE)].[16]

The current density–potential curves are used to highlight the tin blocking effect. They are carried out from the open circuit potential E_{oc} at 5 mV s^{-1} and with an electrode rotation rate of 1000 rev min^{-1}. The galvanostatic polarisation is performed at −10 μA with a still rotation rate.

Characterisation techniques

XPS experiments are performed with a Cameca MAC II analyser operating in the constant analyser energy mode (15 eV pass energy and 1·0 eV energy resolution) and an X-ray source working with the Al K_α anode (mean incidence angle: 57°). The spectrometer energy scale is calibrated in the Auger electron spectroscopy (AES) mode on a clean copper sample with the MVV and LVV Cu transitions at 62·30 and 918·62 eV kinetic energies.[17] After a Shirley type background removal, XPS Cu2p and Sn3d spectra as well as AES CuLVV spectrum are provided to determine qualitatively the film composition.

XPS depth profiles are realised in combination with krypton ion sputtering (differentially pumped Riber ion gun CI-50RB, 3 keV, 57° incidence angle with the sample normal). Using a defocused ion beam to cover the whole sample area exposed to photons, an approximately 0·3 nm min^{-1} sputter rate is deduced by ion

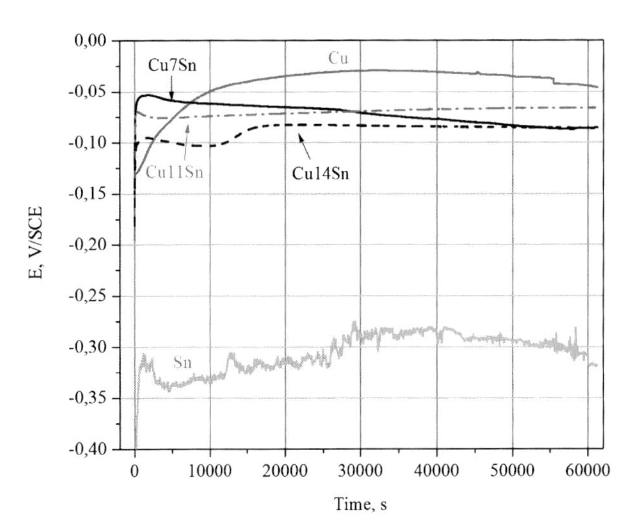

1 Evolution of open circuit potential with immersion time in sulphate electrolyte for copper, tin and $CuSn_7$, $CuSn_{11}$ and $CuSn_{14}$ bronzes

etching on a 41·8 nm Ta_2O_5 reference material certified by nuclear analysis.[18]

Scanning transmission electron microscopy (STEM) experiments are undertaken using a Tecnai F20 with field emission gun operated at 200 kV. The chemical composition is analysed using energy dispersive X-ray. The focused ion beam technique is used to prepare the samples. A platinum protective film is deposed on the corroded sample.

Results and discussion

Effect of tin content on bronze behaviour in sulphate medium

Figure 1 depicts the evolution of the open circuit potential E_{oc} with time, up to 16 h, for the three different Cu–Sn alloys ($CuSn_7$, $CuSn_{11}$ and $CuSn_{14}$) and for pure copper and tin. E_{oc} evolution of pure copper with time exhibits first an increasing curve indicating an interface modification with a passivating film growing on the metal. For a longer immersion time, equilibrium between anodic and cathodic processes is reached, leading to the potential stabilisation.

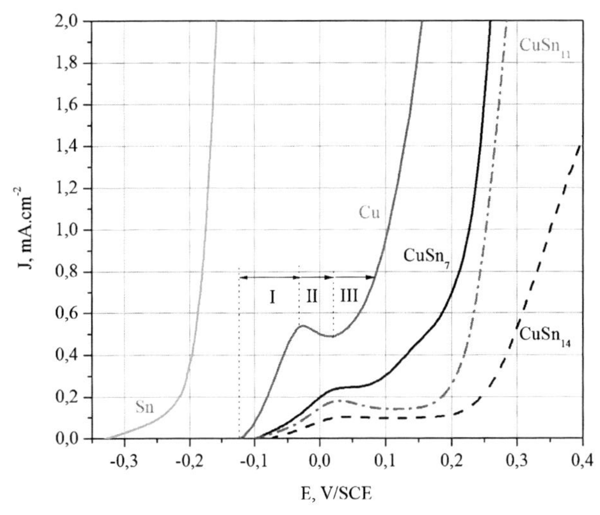

2 Anodic polarisation in sulphate electrolyte at v=5 mV s^{-1} and ω=1000 rev min^{-1} of copper, tin and $CuSn_7$, $CuSn_{11}$ and $CuSn_{14}$ bronzes

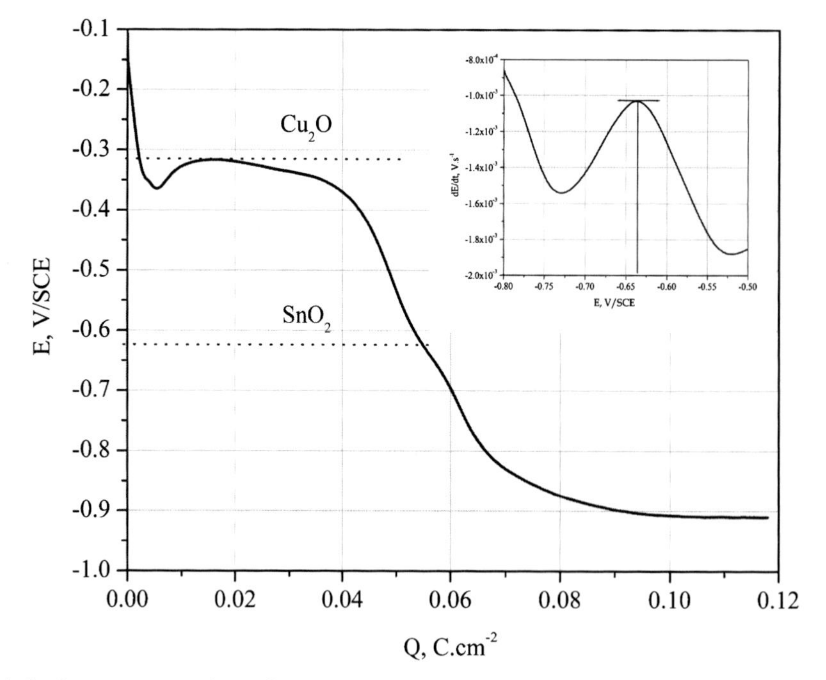

3 **Galvanostatic polarisation curves performed at constant cathodic current of −10 μA on $CuSn_7$ oxidised for 5 min at −0·02 V(SCE) in sulphate electrolyte (derivative curve in insert)**

The E_{oc} curve for pure tin spans a value widely more negative and reveals strong fluctuations of potentials. Such behaviour expresses the dissolution/passivation of the tin electrode.[15] E_{oc} of the Cu–Sn alloys is in the same potential domain as that of copper. The potential shifts first towards a more anodic value, indicating the formation of a protective film at the electrode surface. Then, it shows a progressive decrease (probably caused by an anodic dissolution of copper) until reaching a value between −0·10 and −0·05 V. According to Fig. 1, the higher the tin content, the steeper the slope drop. The increase in tin content should probably hamper the copper dissolution by increasing the thickness of the passive layer. A different film composition or oxide repartition could also explain this trend.

In order to study the influence of tin content on corrosion/passivation processes of bronzes, different specimens have been electrochemically oxidised by applying an anodic potential ramp from the E_{oc} (v=5 mV s^{-1}, ω=1000 rev min^{-1}). The corresponding response in current is reported in Fig. 2. The cathodic pretreatment mentioned before is carried out before any oxidation.

Considering a 30 s death time between the end of the cathodic polarisation and the acquisition of the anodic polarisation, the open circuit potentials of the different specimens available in Fig. 2 are in total agreement with those in Fig. 1.

The tin being a less noble metal than copper, its dissolution occurs at more negative potentials that pure copper. The anodic polarisation curve shows first a soft increase in the current density from E_{oc} up to around −0·20 V. Then, the slope is more pronounced, pointing to a tin dissolution process.

The copper electrode exhibits a different anodic behaviour from that of tin. Three main anodic regions are evidenced. A dissolution ramp (region I) is observed from E_{oc} followed by an anodic peak and a passivation plateau (region II). A sharp increase in the current (current III) occurs in the higher potential range, characteristic of the copper dissolution.

The bronzes behave identically to the copper. The tin alloy content has an effect on the current density of the peak activity. The higher the tin alloy content, the lower the current density of the anodic peak, evidencing a passive layer behaviour. Furthermore, the potential value of the copper dissolution ramp shifts in the noble direction as the Sn content increases, indicating that the tin delays the dissolution process.[19]

Oxides formation, identification and localisation: case of $CuSn_7$ bronze

The aim of this part is to localise and identify the corrosion products formed electrochemically on the $CuSn_7$ bronze and evidenced by galvanostatic polarisation.

In order to accelerate the anodic process of the bronzes, and on the basis of the current–potential curve in Fig. 2, a potential of −0·02 V(SCE) is applied to the $CuSn_7$ electrode for 5 min. This value is taken in the dissolution ramp just before the activity peak.

Indeed, every oxidised compound has a characteristic reduction potential in a given environment. Different copper and tin oxides are reduced at −10 μA and the corresponding potentials are listed in Table 1.

Two steps indicating the presence of two oxides in the passive layer are present on the galvanostatic

Table 1 Preparation method and characteristic reduction potentials of copper and tin oxides in sulphate electrolyte

Nature of the oxides	Preparation method	Reduction potential, V(SCE)
SnO	Composite electrode[20]	−0·58
SnO_2		−0·63
CuO		−0·44
Cu_2O	Chronopotentiometric method[21]	−0·31

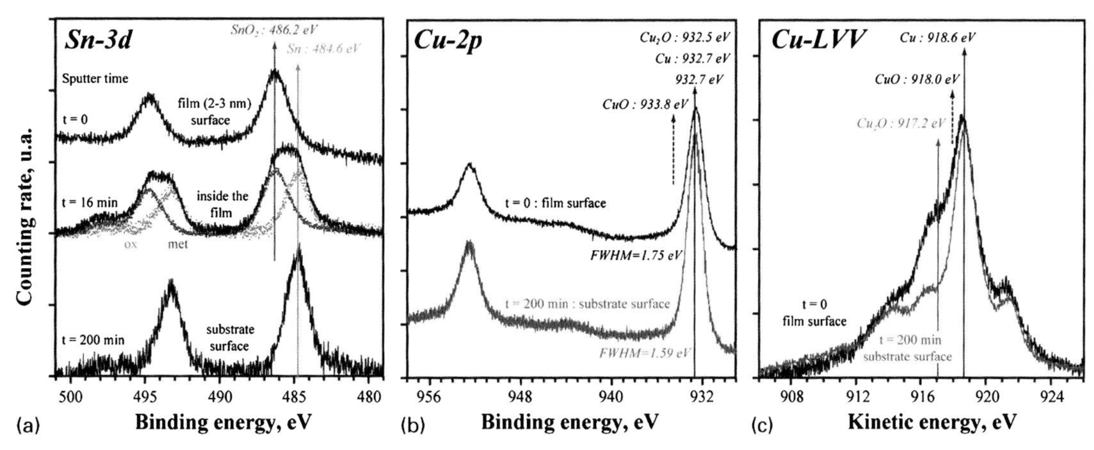

4 ***a*** **Sn-3d,** ***b*** **Cu-2p and** ***c*** **Cu-LVV spectra recorded at different stages of sputter depth profiling of $CuSn_7$ alloy polarised at −0·02 V(SCE) during 5 min**

polarisation of the $CuSn_7$ specimen performed at −10 μA (Fig. 3). The main step is recognisable of the Cu_2O reduction, whereas the potential corresponding to the second step is determined thanks to the derivative curve. The potential of the derivative curve maximum is in the potential range of the SnO_2 reduction potential given in Table 1. In conclusion, the passive layer is constituted of Cu_2O and SnO_2 proving that tin plays a key role in the bronze corrosion.

The localisation and the chemical identification of the different oxides are provided by STEM and XPS examinations. To obtain EDS maps of the different elements, STEM is used preferably to the SEM technique because of the low thickness (a few nanometres) of the film. The same potential of −0·02 V(SCE) is applied on 0·5 mm thick flat samples for 20 min for the STEM experiments and only for 5 min for the XPS investigations. For the latter experiment, the film should be as thin as possible, which is why the polarisation time of the sample is less important.

Sample transfers from the electrochemical cell to the AES–XPS analysis chamber are carefully achieved according to an optimised procedure in order to avoid contacts with the laboratory atmosphere.[22,23]

Initial XPS examination indicates a sample surface mainly constituted by a Sn oxide identified as SnO_2 with a single Sn-$3d_{5/2}$ peak assigned at $486·2\pm0·2$ eV (Fig. 4*a*).[24] At the opposite, the Cu-2p spectrum looks like the metallic copper apart from an enlarged full width at high maximum of 1·75 eV for the $2p_{3/2}$ peak instead of 1·59 eV in pure copper. In fact, this reveals a small Cu_2O spectrum contribution overlapping the Cu substrate one. By another way, a CuO contribution can be excluded on account of the absence of characteristic shake-up satellites in Fig. 4*b*. Because 2p spectra of Cu and Cu_2O are uneasy to differentiate due to their similar morphologies and very close binding energies (ΔE=0·2 eV), it is often recommended with Cu and Cu_2O mixtures to refer to the AES-LVV spectrum induced by photons in which more important shifts between oxides and Cu may be depicted (Fig. 4*c*). Really the important shoulder observed on the left energy side of the LVV peak at a kinetic energy around 917 eV validates qualitatively the existence of Cu_2O at the surface of the film.

Depth profiling of this oxidised film is performed by ion sputtering and the narrowing of the Cu-$2p_{3/2}$ peak as early as the first sputtering sequence proves the fast elimination of a superficial Cu_2O layer. More deeply, the Cu-2p spectrum becomes characteristic of the metallic contribution of the Cu–Sn substrate. The Sn-3d spectra could be decomposed in only two constituents: SnO_2 over the whole depth of the film and Sn proceeding from the underlying substrate. A rough calculation of the signal attenuation of the substrate by the film provides an estimate of the film thickness of about 2–3 nm. Finally, the compositional structure of the present oxidised film can be schematically described by a layer of SnO_2 covered by an ultra thin deposit of Cu_2O.

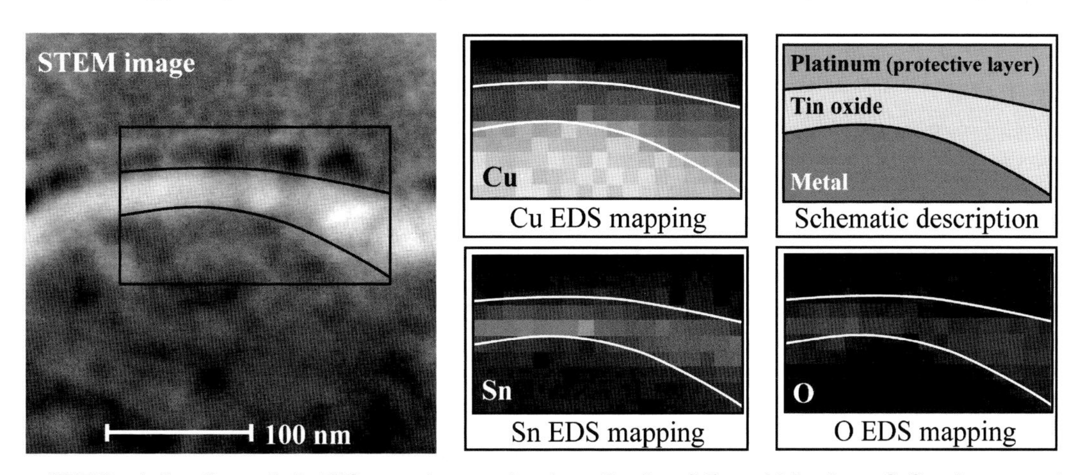

5 Images (STEM) of Cu, Sn and O EDS mappings and schematic description obtained on $CuSn_7$ bronze polarised at −0·02 V(SCE) during 20 min

The STEM image shows that a 15 nm thick layer is formed on the copper–tin sample (Fig. 5). The thickness of the layer is in total agreement with the XPS experiments, considering that the film thickness is proportional to the polarisation time. According to the Cu, Sn and O EDS mappings, this layer is enriched in tin but does not evidence the presence of an external cuprite layer as expected by the galvanostatic polarisation curve (Fig. 3). This can be explained, on the one hand, by the fact that the thickness of the cuprite layer is thinner than that of SnO_2, as observed by the XPS experiments on a 5 min polarised sample and, on the other hand, by the STEM images resolution.

However, these results should be taken cautiously because the presence of copper in the tin enriched layer cannot be excluded, especially when the bulk content is in the range of 5 at-% tin.[16] In fact, other works of the literature indicate the presence of mixed SnO_2 and Cu_2O in the film. Mathis evidences the presence of a mixed oxides layer on samples oxidised at high temperatures[25] and Leyssens *et al.* mention that electrochemical protocol to obtain cuprite on bronzes also generates cassiterite.[26]

Conclusions

The anodic behaviour of Cu–Sn alloys in a 10^{-2} mol L^{-1} sulphate medium consists fundamentally of the formation of a tin enriched passive layer, which becomes more protective with the increase in tin content. According to the XPS and STEM experiments, the passive layer consists mainly in a tin enriched layer, probably mixed with copper, even if the presence of cuprite as an external layer cannot be excluded. This corrosion structure is in agreement with the results found on ancient bronze patina of type I and confirms the results presented before.[12,25]

It is shown that the natural patina on a $CuSn_7$ specimen is able to be reproduced and now, it will be of great importance to perform the methodology developed in this paper on several oxidation potentials and on $CuSn_{11}$ and $CuSn_{14}$. Such a study can provide useful information on the nature, the structure and the stability of the patinas versus the tin content.

A corrosion mechanism will be deduced from these investigations in order to obtain piece of information that will contribute to the development of adequate conservation and restoration treatments.

References

1. L. He, Q. R. Zhao and M. Gao: 'Characterization of corroded bronze Ding from the Yin Ruins of China', *Corros. Sci.*, 2007, **49**, (6), 2534–2546.
2. G. M. Ingo, T. de Caro, C. Riccucci and S. Khosroff: 'Uncommon corrosion phenomena of archaeological bronze alloys', *Appl. Phys. A*, 2006, **83A**, (4), 581–588.
3. M. Wadsak, I. Constantinides, G. Vittiglio, A. Adriaens, K. Janssens, M. Schreiner, F. C. Adams, P. Brunella and M. Wuttmann: 'Multianalytical study of patina formed on archaeological metal objects from Bliesbruck-Reinheim', *Mikrochim. Acta*, 2000, **133**, 159–164.
4. G. P. Cicileo, M. A. Crespo and B. M. Rosales: 'Comparative study of patinas formed on statuary alloys by means of electrochemical and surface analysis techniques', *Corros. Sci.*, 2004, **46**, (4), 929–953.
5. C. Chiavari, K. Rahmouni, H. Takenouti, S. Joiret, P. Vermaut and L. Robbiola: 'Composition and electrochemical properties of natural patinas of outdoor bronze monuments', *Electrochim. Acta*, 2007, **52**, (27), 7760–7769.
6. T. E. Graedel: 'Special issue: copper patina formation', *Corros. Sci.*, 1987, **27**, (7).
7. R. F. Tylecote: 'The effect of soil conditions on the long-term corrosion of buried tin–bronzes and copper', *J. Archaeolog. Sci.*, 1979, **6**, (4), 345–368.
8. H. Otto: 'Über röntgenographisch nachweisbare Bestandteile in Patinaschichten', *Naturwissenschaften*, 1961, **48**, (21), 661–664.
9. I. Constantinides, A. Adriaens and F. Adams: 'Surface characterization of artificial corrosion layers on copper alloy reference materials', *Appl. Surf. Sci.*, 2002, **189**, (1–2), 90–101.
10. E. Bernardi, C. Chiavari, C. Martini and L. Morselli: 'The atmospheric corrosion of quaternary bronzes: an evaluation of the dissolution rate of the alloying elements', *Appl. Phys. A*, 2008, **92A**, (1), 83–89.
11. L. Robbiola, J. M. Blengino and C. Fiaud: 'Morphology and mechanisms of formation of natural patinas on archaeological Cu–Sn alloys', *Corros. Sci.*, 1998, **40**, (12), 2083–2111.
12. P. Piccardo, B. Mille and L. Robbiola: 'Tin and copper oxides in corroded archaeological bronzes', in 'Corrosion of metallic heritage artefacts – investigation, conservation and prediction for long-term behaviour', (ed. P. Dillmann *et al.*), 239–262; 2007, Cambridge, Woodhead Publishing.
13. C. Debiemme-Chouvy, F. Ammeloot and E. M. M. Sutter: 'X-ray photoemission investigation of the corrosion film formed on a polished Cu–13Sn alloy in aerated NaCl solution', *Appl. Surf. Sci.*, 2001, **174**, (1), 55–61.
14. L. Robbiola, T. T. M. Tran, P. Dubot, O. Majerus and K. Rahmouni: 'Characterisation of anodic layers on Cu–10Sn bronze (RDE) in aerated NaCl solution', *Corros. Sci.*, 2008, **50**, (8), 2205–2215.
15. E. Sidot, N. Souissi, L. Bousselmi, E. Triki and L. Robbiola: 'Study of the corrosion behaviour of Cu–10Sn bronze in aerated Na_2SO_4 aqueous solution', *Corros. Sci.*, 2006, **48**, (8), 2241–2257.
16. J. Muller: 'Etude électrochimique et caractérisation des produits de corrosion formés à la surface des bronzes Cu–Sn en milieu sulfate', PhD thesis, Université Paris Est, Thiais, France, 2010.
17. M. P. Seah, G. C. Smith, and M. T. Anthony: 'AES-energy calibration of electron spectrometers. 1. An absolute, traceable energy calibration and the provision of atomic reference line energies', *Surf. Interf. Anal.*, 1990, **15**, (5), 293–308.
18. M. P. Seah, M. W. Holbourn, C. Ortega and J. A. Davies: 'An intercomparison of tantalum pentoxide reference studies', *Nucl. Instrum. Methods Phys. Res. Sect. B*, 1988, **30B**, (2), 128–139.
19. B. G. Ateya, J. D. Fritz and H. W. Pickering: 'Kinetics of dealloying of a copper-5 atomic percent gold alloy', *J. Electrochem. Soc.*, 1997, **144**, (8), 2606–2613.
20. B. Laik, F. Gessier, F. Mercier, P. Trocellier, A. Chausse and R. Messina: 'Influence of lithium salts on the behaviour of a petroleum coke in organic carbonate solutions', *Electrochim. Acta*, 1998, **44**, (10), 1667–1676.
21. B. Laik, P. Poizot and J. M. Tarascon: 'The electrochemical quartz crystal microbalance as a means for studying the reactivity of Cu_2O toward lithium', *J. Electrochem. Soc.*, 2002, **149**, (3), A251–A255.
22. F. Basile, J. Bergner, C. Bombart, B. Rondot, P. Le Guevel and G. Lorang: 'Electrochemical and analytical (XPS and AES) study of passive layers formed on Fe–Ni alloys in borate solutions', Proc. 8th European Conf. on 'Applications of surface and interface analysis', Seville, Spain, Philippe Dillmann October 1999, 154–157.
23. M. Bouttemy: 'Etude des mécanismes de formation et de croissance des films passifs formés sur les alliages Fe–Ni et Fe–Cr', PhD thesis, Université Paris Sud – Paris XI Paris, Paris, France, 2006.
24. P. Keller and H. H. Strehblow: 'XPS-studies on the formation of passive layers on Sn and CuSn alloys in 0·1 m KOH', *Z. Phys. Chem.*, 2005, **219**, (11), 1481–1488.
25. F. Mathis: 'Croissance et propriétés des couches d'oxydation et des patines à la surface d'alliages cuivreux d'intérêt archéologique ou artistique', PhD thesis, Université d'Orsay (Paris 11), Paris, France, 2005.
26. K. Leyssens, A. Adriaens, C. Degrigny and E. Pantos: 'Evaluation of corrosion potential measurements, as a means to monitor the storage and stabilization process of archaeological copper-based artifacts', *Anal. Chem.*, 2006, **78**, 2794–2801.

Improvement of corrosion stability of patinated bronze

H. Otmačić Ćurković[1], T. Kosec*[2], A. Legat[2] and E. Stupnišek-Lisac[1]

Bronze surfaces suffer from corrosion processes if they are exposed outdoors. They are affected by the presence of different aggressive species, the alloy composition and surface preparation. Bronze can be exposed outdoors non-protected or can be pretreated in different ways. Artists use chemical patinations in order to achieve visual effects. In the present study, different finishes were tested on three different patinas: green nitrate, green chloride and an electrochemically prepared patina. The tested finishes were: imidazole and benzotriazole type inhibitors in either ethanol or Paraloid B44, and Carnauba wax as a representative of waxes. The effectiveness of the applied finishes was examined by electrochemical impedance spectroscopy in a solution simulating urban acid rain. Aging experiments were performed in a climatic chamber and a salt spray chamber in order to combine different exposure conditions, the presence of sulphur dioxide, high humidity and a mist of chloride ions. After immersion in an urban acid rain solution, acidified to pH 5, the different patinas were investigated by SEM/EDX and Raman spectroscopy.

Keywords: Bronze, Patina, Benzotriazole, Tolyl metyl imidazole, EIS, Raman spectroscopy

This paper is part of a special issue on corrosion of archaeological and heritage artefacts

Introduction

When exposed to an outdoor environment, bronze sculptures undergo various corrosion processes owing to the influence of humidity, rain and various pollutants such as SO_2, NO_x, CO_2.[1,2] Under such conditions the corrosion products (patina) that can be present on the surface of such sculptures are also prone to dissolution, and can therefore offer little protection to the underlying bronze.

Because of the artistic and archaeological value of patina, usually both the bronze and its patina need to be preserved. In order to satisfy the requirements for artefacts belonging to the cultural heritage (i.e. unaltered appearance, removable protective system), only some corrosion protection methods, such as waxes, varnishes and lacquers and corrosion inhibitors, may be applied, often in combination.[2–6] In this work, the protective effect of two azole based corrosion inhibitors, either by themselves or together with an acrylic coating, has been studied. One of these inhibitors, which is frequently used in conservation practice,[2–9] is benzotriazole (BTAH), whereas the second is 4-methyl-1-(*p*-tolyl) imidazole (TMI), which is an environmentally friendly compound that has also proved to be a good corrosion inhibitor for copper and its alloys in various aggressive environments.[10–12] In the authors' previous work, an evaluation was performed on the toxicity of a series of 4-methyl imidazole derivatives on the biological system for the treatment of waste water. This evaluation showed that the tested inhibitors have a relatively low toxicity.[13]

The effectiveness of a protection system can be validated by monitoring its stability and the degree of protection which it provides when exposed to corrosive media. Tests are commonly performed in a climatic chamber, or in a solution which simulates the electrolyte that forms when moisture condenses, or in the case of rain in which O_2 and pollutant gases are dissolved. Electrochemical techniques are applied in order to determine the corrosion rate in such conditions.

In this work, the protective effect of corrosion inhibitors and an acrylic coating is examined by means of electrochemical impedance spectroscopy (EIS) measurements, which were performed after different immersion times in an electrolyte simulating rain in a polluted urban environment. The results obtained are compared to those obtained in the climatic chamber. Electrochemical impedance spectroscopy is frequently used for the verification of the effectiveness and stability of protective systems.[2,3,6,7,10–12]

In the case of this research, the patinas were either obtained by the chemical methods that are used by artists in order to produce an antique appearance of bronze, or by electrochemical procedures which can provide a patina similar to that which forms spontaneously on bronze artefacts.

[1]Faculty of Chemical Engineering and Technology, University of Zagreb, Savska 16, 10000 Zagreb, Croatia
[2]Slovenian National Building and Civil Engineering Institute, Dimičeva 12, 1000 Ljubljana, Slovenia

*Corresponding author, email tadeja.kosec@zag.si

Published by Maney on behalf of the Institute
Received 15 December 2009; accepted 25 March 2010
DOI 10.1179/147842210X12710800383684

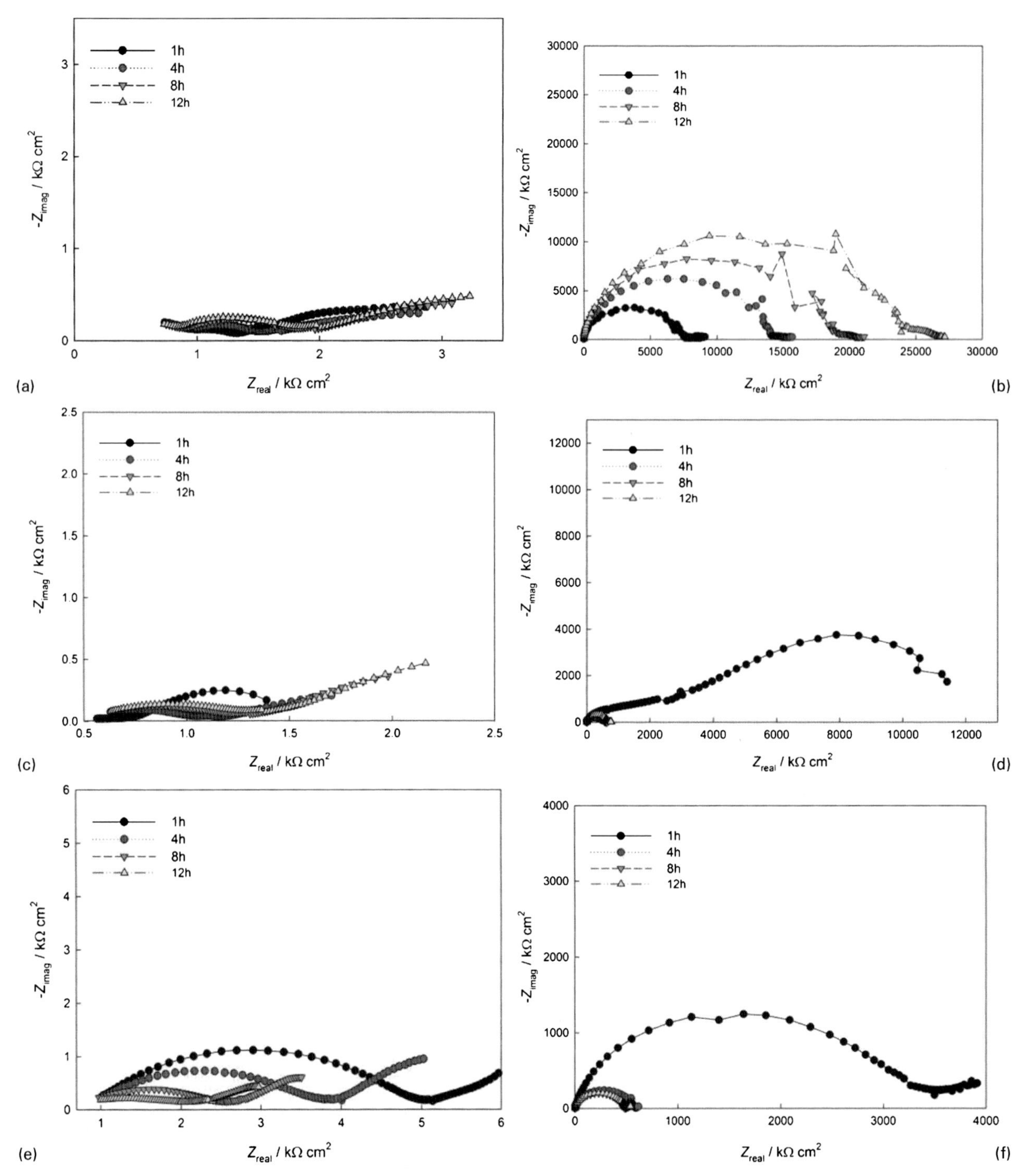

a green nitrate patina; *b* Carnauba wax; *c* BTAH in ethanol; *d* BTAH in Paraloid B44; *e* TMI in ethanol; *f* TMI in Paraloid B44

1 Time dependence of EIS spectra for nitrate green patinated bronze, *a* unprotected and *b–f* protected

Experimental

Sample preparation

The bronze samples were cast at a privately owned foundry, and consisted of 87 wt-% copper, 5·5 wt-% tin, 3·2% of zinc, 0·6% of Al and 3·2% of impurities, as analysed by an X-ray Fluorescence analyser. After casting, the bronze was sand blasted. The 100× 100×5 mm samples, which were exposed in a climatic chamber, were patinated, and suitably protected by different protection methods. The samples for the electrochemical tests and the immersion tests were cut, in the shape of discs or rectangles, from plates, and abraded with 800 and 1000 grid SiC paper. The samples were cleaned in distilled water and then well dried.

Three different patinas were studied. Two patinas were obtained chemically. These two patinas had a green colour. The third patina was an electrochemical patina, which resembled the naturally formed patina which occurs when bronze is exposed to an urban atmosphere.

A two step procedure was used in order to obtain the chemically formed patinas. In both cases, a brown patina was first formed on the bronze, by brushing the hot surface with a 3% solution of K_2S. In the second step, a chloride type green patina was formed by brushing the previously formed hot brown patinated surface with a solution of NH_4HCO_3 and NH_4Cl

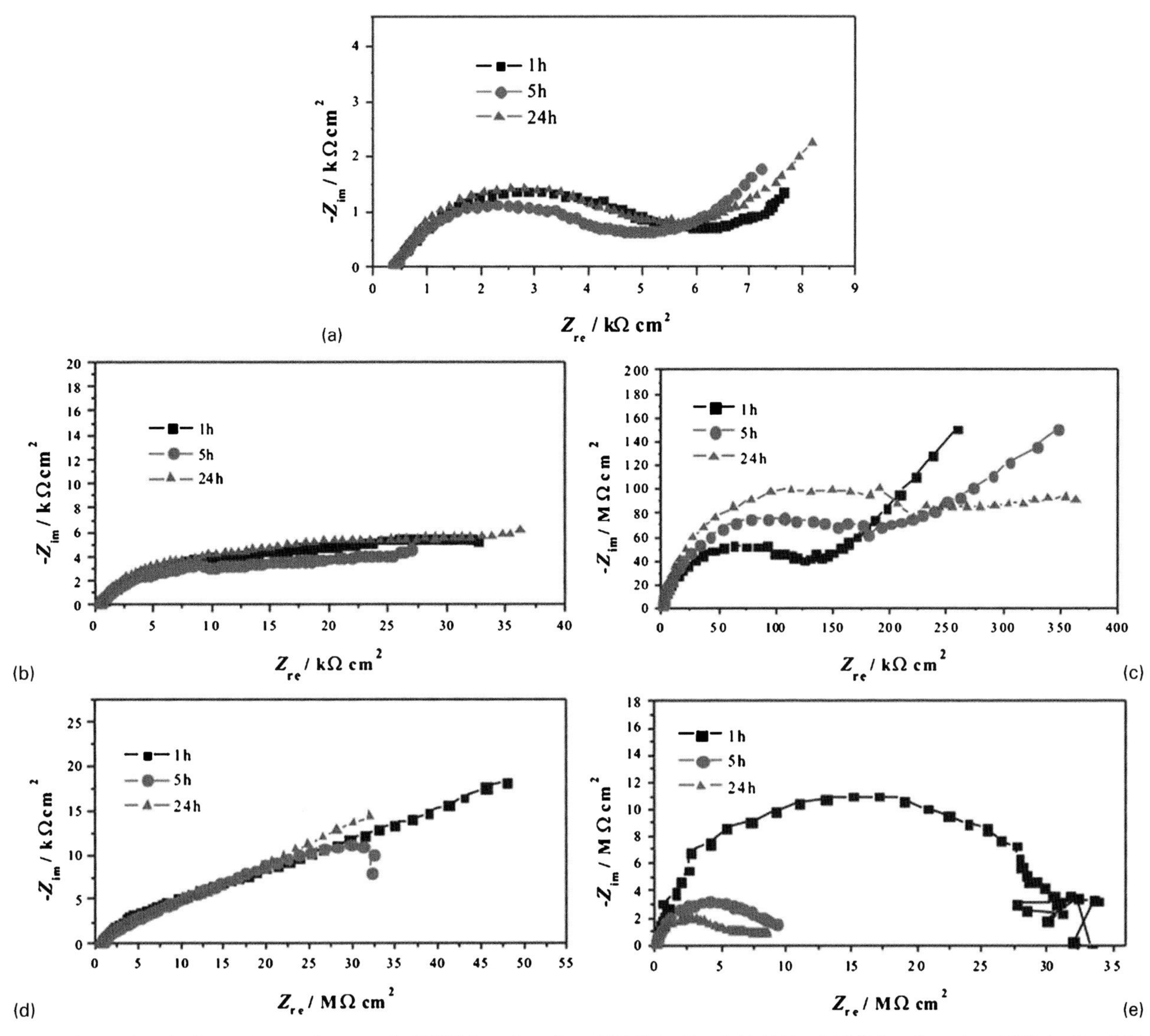

a electrochemically patinated bronze; *b* BTAH in ethanol; *c* BTAH in Paraloid B44; *d* TMI in ethanol; *e* TMI in Paraloid B44

2 Time dependence of EIS spectra for nitrate green patinated bronze, *a* unprotected and *b–e* protected

(30 g L^{-1} each), whereas the nitrate patina was achieved by brushing the previously formed hot brown patinated surface with a 3% solution of $Cu(NO_3)_2$.

The electrochemical patina was synthesised in a solution of 0·2 g L^{-1} Na_2SO_4, 0·2 g L^{-1} $NaHCO_3$ and 0·2 g L^{-1} $NaNO_3$ after a stable open circuit potential (OCP) had been reached under potential control: 60 s at −0·20 V versus OCP, 48 h at +0·14 V versus OCP, and a further 48 h at +0·12 V versus OCP. A similar procedure was used to prepare the electrochemical patina in a sulphate/carbonate solution.[10,11]

The following finishes were selected to be tested as a top coat over the prepared patinas: TMI inhibitor, and BTAH in an ethanol solution or in an ethyl acetate solution with an ethylmetacrylate/methyl metacrylate copolymer (Paraloid B44) as a 3% solution of the inhibitor. Carnauba wax was also tested. It was melted and mixed with White spirit. Different finishes were applied over the patinated bronze using a brush.

Exposure and testing conditions

The exposure pattern for the climatic exposure experiments was as follows: 3 week industrial chamber (Kohler, model HK 430, Germany),[14] 8 week humidity chamber,[15] 2 days salt spray chamber[16] (Votsch - model VSC/KTW 1000) and 10-day humidity chamber.

The small samples with different patinas were immersed in a solution of 0·2 g L^{-1} Na_2SO_4, 0·2 g L^{-1} $NaHCO_3$ and 0·2 g L^{-1} $NaNO_3$ (simulated acid rain) acidified to pH 5 with diluted H_2SO_4 in order to simulate urban rain for 35 days. Subsequently the samples were rinsed with distilled water and dried. The test samples were then examined.

Electrochemical impedance spectroscopy measurements were conducted in the same solution of urban acid rain in the frequency range between 65 kHz and 5 mHz at OCP, and the amplitude of the perturbing signal was 10 mV. Ten points were collected per decade.

SEM analysis and Raman spectroscopy

Scanning electron micoscopy (SEM) was performed on a low vacuum JEOL 5500 LV scanning electron microscope (JEOL, Japan), equipped with Oxford Inca (EDX) energy dispersive spectroscopy (Oxford Instrument Analytical, UK), in order to examine the surface products which were formed using an accelerating voltage of 20 kV.

The Raman spectra were recorded by means of a Horiba Yvon HR800 spectrometer. The samples were

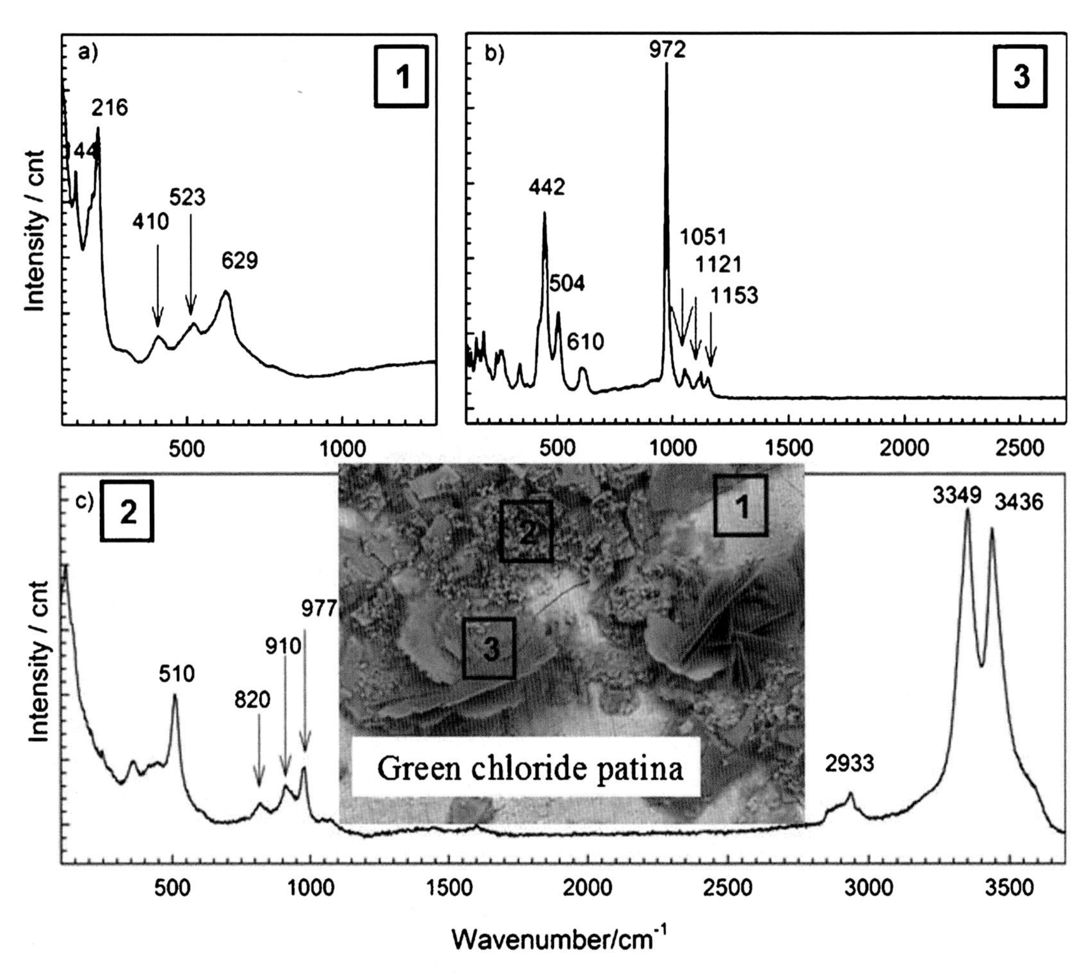

3 Image (SEM) of green chloride patina and Raman spectra for corresponding *a* area no 1, *b* area no 3 and *c* area no 2

irradiated with a green laser at λ=514 nm. The laser power was set at 14 mW, at a magnification of 100 times. The scanning range was 50–3700 cm^{-1}, each acquisition with an accumulation time of 20–35 s, depending on the number of counts.

Results and discussion

Electrochemical impedance spectroscopy

The EIS measurements were performed in the solution which simulated acid rain in an urban environment. Figure 1*a* shows the impedance evolution over time for unprotected bronze with a chemically obtained nitrate patina. Owing to the increase in the impedance over time it can be assumed that less stable patina compounds are dissolved or transformed into more stable compounds. The chloride patina exhibited similar behaviour. The impedance of both chemical patinas protected by Carnauba wax was significantly greater than that of the unprotected samples. However, a ten times higher impedance was observed in the case of the wax treated nitrate patina (Fig 1*b*), which also tended to increase over time. Protection by means of an ethanol solution of the studied corrosion inhibitors [benzotriazole and 4-methyl-1-(*p*-tolyl) imidazole] does not seem to be effective, except in the case of the nitrate patina protected by TMI (Fig. 1*e*). However, after long immersion times, the impedance decreases to values which are similar to those observed in the case of the unprotected sample. Protection of the investigated corrosion inhibitors by means of a Polaroid B44 solution (Fig. 1*d* and *f*) was found to be very effective, although this impedance tended to decrease over time. Its value was very high even after 24 h of immersion in simulated acid rain.

Measurements were also performed on the electrochemically patinated bronze samples. As can be seen from Fig. 2*a*, the impedance decreased slightly after a few hours of immersion, but after 24 h it was practically the same as immediately after immersion. Its value was higher than that observed in the case of the chemically patinated samples, which suggests that the electrochemically obtained patina provides the underlying bronze with better protection. The EIS spectra of patinated bronze samples protected by the investigated inhibitors dissolved in ethanol are presented in Fig. 2*b* and *d*. It may be observed that both inhibitors increased the impedance of the patinated bronze. In the case of the BTAH treated sample, a slight decrease in impedance was observed after a few hours of immersion, but later on it became almost the same as at the beginning of the immersion. In the case of TMI, impedance decreased slightly during the first few hours of immersion, but later on it remained constant. From the results of the EIS measurements, it can therefore be concluded that both inhibitors exhibit similar efficiency on patinated bronze.

Impedance measurements were also performed on samples protected with the inhibitor solutions in Polaroid B44 (Fig. 2*c* and *e*). Although the use of either of the inhibitors resulted in a much increased bronze impedance, the inhibiting effect of BTAH was found to

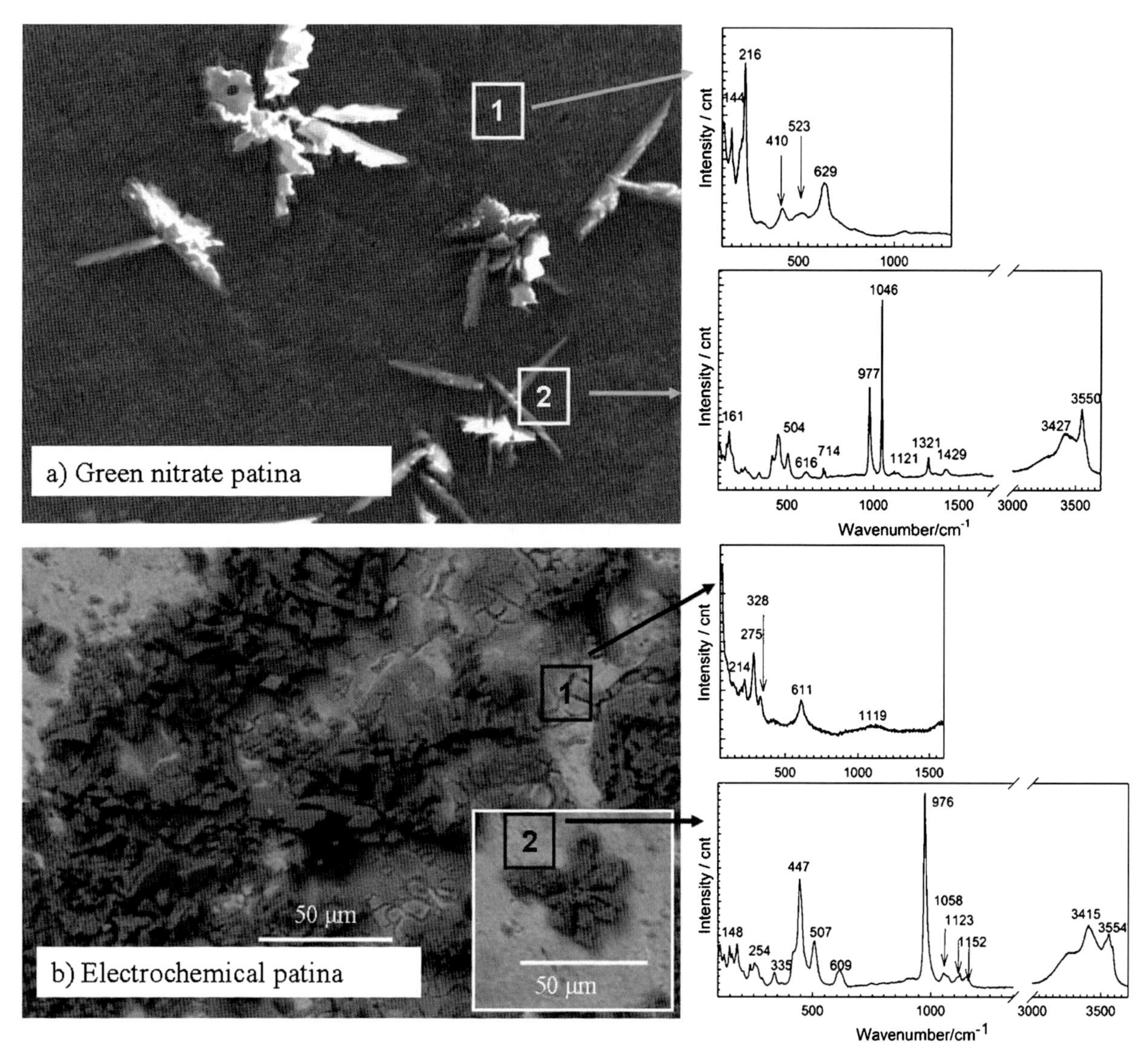

4 Image (SEM) of green nitrate and electrochemical patinas and corresponding Raman spectra

be 10 times greater than that of TMI. Also, the impedance of the BTAH treated sample slightly increased over time, whereas that of the TMI treated sample decreased. The largest decrease in impedance was observed during the first few hours of immersion. Still, after 24 h of immersion, the impedance of the TMI treated sample was 1000 times greater than that of the unprotected patinated sample.

The investigated protection methods showed a better inhibiting effect on the electrochemically patinated bronze than on the chemically patinated bronze.

Immersion in urban acid rain solution

The green chloride patina, the green nitrate patina and the electrochemical patina were immersed in urban acid rain, pH 5 in order to study the corrosion products that formed during the 35 day exposure. The SEM images with Raman spectra are presented in Figs. 3 and 4.

Three different corrosion products were formed on the green chloride patina, and are presented and marked in Fig. 3: fine round crystals which cover the surface of the bronze (area no. 1), round crystals that are

Table 1 EDX analysis of green chloride patina, green nitrate and electrochemical patina after 35 day immersion in simulated urban rain

Material, at-%	Cu	Sn	Zn	S	Cl	O	N	C
Green chloride patina								
No. 1	87·32	0	0	0·69	0·91	10·83	1·72	0
No. 2	9·97	0·02	0·33	2·57	0·28	75·30	7·40	4·13
No. 3	21·32	0·05	0·82	0·65	4·02	58·14	8·41	6·60
Green nitrate patina								
No. 1	51·14	0	0	0·14	0	37·27	8·19	4·13
No. 2	20·61	0·02	0·07	2·67	0	62·81	8·53	5·30
Electrochemical patina								
No. 1	20·57	6·25	0	1·0	0	56·07	6·65	9·48
No. 2	35·35	0·31	0	4·43	0	52·0	4·01	3·90

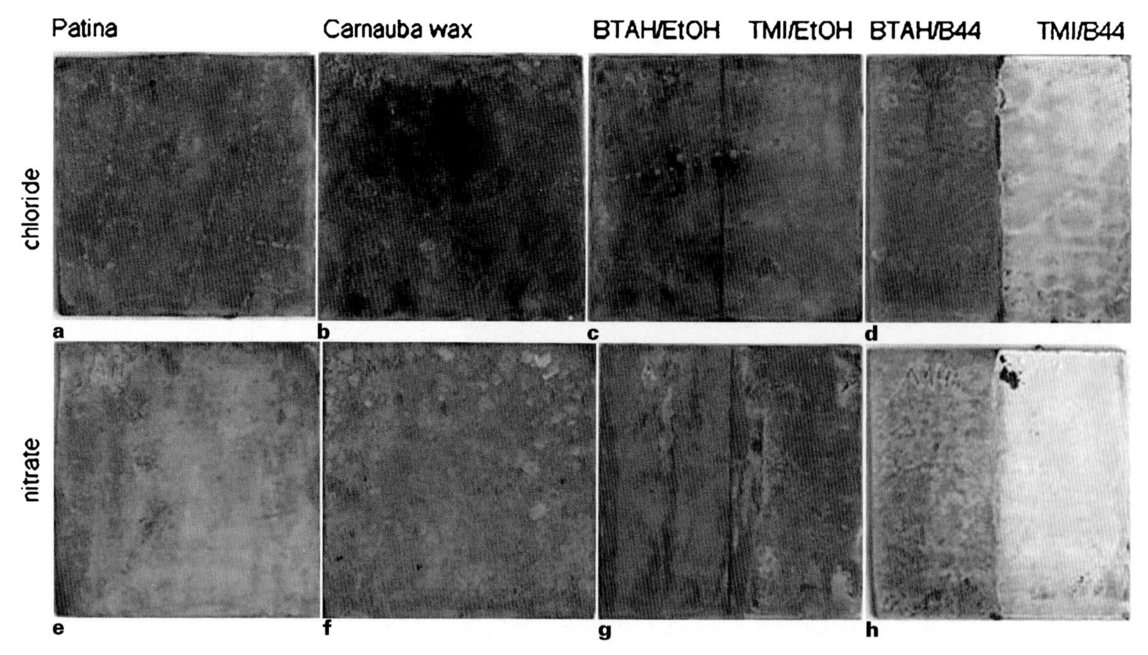

5 Pictures of chemically patinated and protected bronze surfaces, after 12 weeks of exposure to climatic chamber

distributed at different places (area no. 2), and flower-like crystals (area no. 3).

EDX analysis of area no. 1 showed the presence of copper and oxygen (Table 1). Sulphur, chlorine and nitrogen are present in trace amounts. The presence of cuprous or cupric oxide was assumed. The Raman spectrum at this spot consists of two main features: low frequency narrow bands at 144 and 216 cm^{-1}, and three broad bands positioned at 410, 523 and 629 cm^{-1}. The spectrum confirmed the presence of cuprite, Cu_2O as already reported.[10,17] Area no. 2 denotes agglomerated small round crystals, whose EDX analysis showed the presence of chlorine, nitrogen and carbon next to copper and oxygen (Table 1). The Raman spectrum reveals several characteristic bands at 510, 820, 910, 977 cm^{-1} and two bands at 3349 and 3436 cm^{-1}. This spectrum can be identified as atacamite $Cu_2(OH)_3Cl$.[10,18,19] A spectrum characteristic for copper carbonates and nitrates was not observed, so that it may be assumed that they were present in the amorphous phase. At area no. 3, flower-like crystals were found. The copper sulphate mineral is likely to be present owing to presence of sulphur, the lower atomic fraction of copper, and the higher fraction of oxygen in comparison with the cuprite area (Table 1). The Raman spectra confirmed the presence of brochantite $Cu_4(SO_4)(OH)_6$ with main characteristic bands at 974 cm^{-1}, which indicate the presence of SO_4^{2-}, and bands owing to bending features at 442, 504 and 610 cm^{-1}, as noted already in the literature.[10,18,20]

Two different areas were found on the green nitrate patina after 35 day immersion (Fig. 4*a*), as confirmed by EDX analysis. The Raman spectrum at area no. 1 revealed the presence of cuprite over a large part of the area. At area no. 2 the Raman spectrum showed the following characteristic bands: bands at 412, 446 and 504 cm^{-1}, very strong bands at 976 and 1046 cm^{-1}, weak bands at 1123, 1321 and 1429 cm^{-1}, and two wide and strong bands at 3427 and 3550 cm^{-1}. The spectrum reveals the presence of the nitrate mineral ruaite.[18] Also, brochantite is a possible corrosion product.[10,18–20] It might have replaced the more soluble nitrate salts after immersion in the urban rain solution.

Two different corrosion products were found on the electrochemically formed patina (Fig. 4*b*). EDX analysis at area no. 1 showed a fairly large amount of tin, so the formation of SnO_2 was assumed, and since the main bands belonging to tin oxides are close to those belonging to cuprite, a mixture of both oxides can be proposed. EDX analysis showed the presence of sulphur at area no. 2, where well defined crystals were found. The Raman spectrum showed characteristic bands at 148, 176, 233, 254 and 335 cm^{-1}, a triplet at 446·5, 507 and 609 cm^{-1}, a very strong band at 976 cm^{-1}, weak bands at 1058, 1123 and 1152 cm^{-1}, and broad bands at 3415 and 3554 cm^{-1}. The spectrum confirms the presence of brochanitite.[10,19]

Climatic chamber exposure

Figure 5 shows images of the patinated and protected bronze patina surfaces after exposure to the climatic chamber for 12 weeks. The green chloride patina (Fig. 5*a*) did not change a great deal after the exposure. The colour deepened and became intense. When Carnauba wax was applied to the surface of green patinated sample, the wax moistened the patina, which became dark green. After exposure, several patches of a green corrosion product appeared over 10% of the surface (Fig. 5*b*). Also, the BTAH and TMI inhibitors in an ethanol solution were applied to the green chloride patinated sample, and after 12 weeks of exposure the green colour remained, but it was observed that the intensity of the green colour on the left side of the surface, where BTAH had been applied, was somewhat reduced (Fig. 5*c*). A shiny appearance was observed after applying both inhibitors in Paraloid B44 over the green chloride patina. After the 12-week exposure, the surface was changed to a whitish colour (Fig. 5*d*). At some areas the top coat started to delaminate.

A fresh green nitrate patina has a blue touch to its basic green colour. After exposure, some patches of the surface changed to a dark brown colour (Fig. 5*e*). When

Carnauba wax was applied over the surface, a very nice shade of patina colour was retained, in comparison with the green chloride patina. After 12 weeks, colour had deepened (Fig. 5*f*). Visual examination showed that the Carnauba wax was not removed from the surface. When the two inhibitors were applied to green nitrate patina, it was shown that TMI in EtOH induce change in the visual appearance of the patina in comparison with BTAH/EtOH, where it remained unchanged (Fig. 5*g*). Similarly as in the case of the green chloride patina, Paraloid B44 is not a suitable solution for application as a long term protection for patinated bronze (Fig. 5*h*).

Results obtained in this work point out that the ability of an inhibitor to protect patinated bronze depends on the composition of patina. From the electrochemical measurements, it can be concluded that for some types of patina, the environmentally friendly corrosion inhibitor TMI can provide protection similar to that of BTAH, which is presently commonly used in conservation practice.

Conclusions

Electrochemical experiments, immersion tests over a period of 35 days, with spectroscopic studies and exposure in a climatic chamber, were performed on different chemically and electrochemically obtained patinas on bronze samples. The different tested finishes were as follows: TMI and BTAH inhibitors in either an ethanol or ethyl-acetate solution – Paraloid B44, and Carnauba wax as a typical representative of waxes. The results of electrochemical impedance spectroscopy show that neither of the chemically prepared patinas could protect the underlying bronze from further corrosion processes. The electrochemical patina, that resembles a natural patina, is more protective due to the tin and cuprous oxides which form over time. The inhibitors dissolved in an ethanol solution do not seem to protect the underlying patinas in the case of green chloride and green nitrate patina. The protective behaviour of TMI and BTAH inhibitors in an ethanol solution is increased on electrochemical patina. The electrochemical results of the two inhibitors in Paraloid B44 showed their good protective ability, but exposure to climatic chamber conditions proved their inadequacy for use on chemically prepared bronzes. From the results of electrochemical measurements it can be concluded that, in the case of some types of patina, the environmental friendly corrosion inhibitor TMI can be used as a substitute for BTAH.

The corrosion products that formed during 35 days of exposure to urban acid rain consisted of cuprite, atacamite and brochantite on the chloride patina, of cuprite and brochantite on the nitrate type patina, and brochantite, tin and cuprous oxides on the electrochemically formed patina.

The electrochemical results and the results of the exposure of different patinas in the climatic chamber point to the adequacy of the use of Carnauba wax.

Acknowledgements

The financial support of the Ministry of Science and Technology of the Republic of Croatia, under Project 125-1252973-2572, and of the Slovenian Research Agency, under Grant No. Z2-2298, is hereby gratefully acknowledged.

References

1. M. Tullmin and P. R. Roberge: 'Uhlig's corrosion handbook', 305–321, 2000, New York, John Wiley & Sons.
2. G. Bierwagen, T. J. Shedolsky and K. Stanek: *Prog. Org. Coat.*, 2003, **48**, 289–296.
3. P. Letardi: Proc. Metal 2004, Canberra, Australia, October 2004, National Museum of Australia, 379.
4. V. Argyropouls, E. Angelini and C. Degrigny: Proc. Metal 2004, Canberra Australia, October 2004, National Museum of Australia, 43.
5. E. Franceschi, P. Letardi and G. Luciano: *J. Cult. Herit.*, 2006, **7**, 166–170.
6. A. Galtayries, A. Mongiatti, P. Marcus and C. Chiavari: in 'Corrosion of metallic heritage artefacts', (ed. P. Dillmann *et al.*), 335–351; 2007, Cambridge, EFC Woodhead Publishers.
7. Rahmouni, H. Takeonutti, H. Hajjjaji, A. Srhiri and L. Robbiola: *Electrochim. Acta*, 2009, **54**, 5206–5215.
8. S. Golfomitsou and J. F. Merkel: Proc. Metal 2004, Canberra, Australia, October 2004, National Museum of Australia, 344.
9. L. A. Zycherman and N. F. Veloz: *J. Am. Inst. Conservat.*, 1979, **19**, 24–33
10. K. Marušić, H. Otmačić-Ćurković, Š. Horvat-Kurbegović, H. Takenouti and E. Stupnišek-Lisac: *Electrochim. Acta*, 2009, **54**, 7106–7113.
11. L. Muresan, S. Varvara, E. Stupnišek-Lisac, H. Otmačić, K. Marušić, S. Horvat-Kurbegović, L. Robbiola, K. Rahmouni and H. Takenouti: *Electrochim. Acta*, 2007, **52**, 7770–7779.
12. H. Otmačić and E. Stupnišek-Lisac: *Electrochim. Acta*, 2003, **48**, 985–991.
13. E. Stupnišek-Lisac, A. Lončarić Božić and I. Cafuk: *Corrosion*, 1998, **54**, 713–720.
14. 'Metallic and other non-organic coatings – sulphur dioxide test with general condensation of moisture (ISO 6988:985)', SIST EN ISO 6988:1999.
15. 'Paints and varnishes – determination of resistance to humidity – Part 2: procedure for exposing test specimens in condensation-water atmospheres', SIST EN ISO 6270-2 ATH.
16. 'Corrosion tests in artificial atmospheres – salt spray tests', SIST EN ISO 9227 NNS:2006.
17. G. Niuara: *Electrochim. Acta*, 2000, **45**, 3507–3519.
18. V. Hayez, V. Costa, J. Guillaume, H. Terryn and A. Hubin: *Analyst*, 2005, **130**, 550–556.
19. L. Frost, P. Leverett, P. A. Williams, M. L. Weier and K. L. Erickson: *J. Raman Spectrosc.* 2004, **35**, 991–996.
20. V. Hayez, T. Segato, A. Hubin and H. Terryn: *J. Raman. Spectrosc.*, 2006, **37**, 1211–1220.

Silver artefacts: plasma deposition of SiO_x protective layers and tarnishing evolution assessment

E. Angelini*[1], S. Grassini[1] and M. Parvis[2]

Silver artifacts suffer tarnishing when exposed to a sulphur containg atmosphere. Plasma deposited SiO_x thin films are proposed for the protection of silver artefacts, owing to their optical transparency and high barrier properties against vapours. The protective effectiveness of the SiO_x films was assessed by means of electrochemical impedance measurements performed in an Na_2S solution on a set of coated silver based alloy samples, and by submitting another set of samples to a tarnishing test in presence of H_2S vapours. The experimental findings reveal that the SiO_x deposition, performed in RF plasma fed with a tetraetoxysilane/oxygen/argon mixture, produces layers with excellent barrier effects against the aggressive agents. The protective effectiveness increases if the deposition is performed at increasing input powers and at decreasing tetraethoxysilane/oxygen ratios in the feeding gas. The tarnishing evolution onto the coated surface was assessed at the microscopical and macroscopical level by means of field emission scanning electron microscopy and by an easy to use diagnostic tool based on digital photography and image processing.

Keywords: Silver, SiO_x protective films, Tarnishing, Electrochemical impedance, Cultural heritage

This paper is part of a special issue on corrosion of archaeological and heritage artefacts

Introduction

Silver, either as pure metal or in association with other metals such as copper or gold, has been used since ancient times in the fabrication of decorative and functional objects owing to its working properties and pleasant color and shine.

Although highly lustrous when polished, silver and silver based alloys gradually darken and become less shiny when exposed to the atmosphere for a long time. The well known tarnishing reaction, indeed, starts almost immediately, due to the interaction with sulphur containing compounds present in the environment.[1] The main compound responsible of silver tarnishing is hydrogen sulphide (H_2S); other organic sulphides like carbonyl sulphide,[2] however, even if present in lower concentrations, can react more rapidly with silver and contribute to the degradation process.

The tarnishing reaction takes place through a variety of mechanisms according to the environmental conditions, temperature and relative humidity, together with other corrosion processes that may take place at the same time.[3,4] However, the main product of silver tarnishing is silver sulphide (Ag_2S), whose formation proceeds according to the following electrochemical reaction:

$$2Ag + S^{2-} \rightarrow Ag_2S + 2e^-$$

For this reason, historic–artistic artefacts are subjected to a gradual discoloration and loss of polished finish; the metal surface starts darkening to brown before ending up as a very dark grey or black with a slight sheen.

The removal of the tarnishing layer may be performed with different techniques, mechanical, chemical and electrochemical. By means of mechanical polishing and/or commercial liquid cleaners, which often contain abrasive compounds, a thin silver layer along with the tarnishing may be inadvertantly removed. Cathodic polarisation in sodium sesquicarbonate is preferable because it is less aggressive and effective also on gilt silver artefacts.[5] In particular on these artefacts a preliminary cathodic polarisation step is followed by an anodic polarisation that removes the reduced silver from the surface of the artefacts.[6]

After the cleaning step, the best way to avoid or delay further degradation is, according to International Council of Museums (ICOM-ICC), preventive conservation, i.e. indirect measures or actions that do not interfere with the artefacts.[7] In the specific case of metallic artefacts measures as reduction of environmental relative humidity, reduction of the concentration of environmental pollutants, employment of vapour phase corrosion inhibitors, as morpholine-n-methylmorpho-

[1]Department of Materials Science and Chemical Engineering, Politecnico di Torino, Italy
[2]Department of Electronics, Politecnico di Torino, Italy

*Corresponding author, email emma.angelini@polito.it

Published by Maney on behalf of the Institute
Received 11 January 2010; accepted 5 July 2010
DOI 10.1179/147842210X12767807773484

line-camphor type, oil-soluble chlorophyl, and sodium copper chlorophylline[8,9] may be undertaken. In many cases, however, it is difficult to maintain the artefacts under proper environmental conditions, therefore, in order to prevent further tarnishing, a direct application of commercially available inhibitors or lacquerers on the silver artefacts is performed. The protective effectiveness of these treatments lasts few months so the need to repeat cleaning and protecting procedures can increase the damage. Moreover, most of the available materials, routinely used by conservators and restorers, are unsafe and environmentally dangerous.[10]

Summarising, there is a continuous search for new and improved coatings to solve the neverending problem of cleaning, protecting and preventing further damages of restored silver artifacts and in general for metallic artefacts.[11,12] As a matter of fact, the ethics of restorers and conservators requires the maintenance of the aesthetic appearance and the integrity of the object, without removing or modifying parts of it. To satisfy curators requirements, protective coatings must be transparent, with an high chemical stability, and reversible in order to allow the recovery of the initial state of the object, if necessary. At the same time, protective coatings should provide a continuous barrier against aggressive agents; so any defect can become a focal point when a localised corrosion attack can start.

For these reasons, a satisfactory conservation approach requires, on one hand, the development of coatings with high protective effectiveness and, on the other hand, a constant monitoring, at macroscopic level, of their stability and of the degradation phenomena occurring onto the artifact's surface.

As shown by the authors in previous studies,[13,14] plasma deposited SiO_x thin films suit satisfactorily this application and are proposed in this study to provide long term protection against sulphur containing atmosphere to silver based alloys. The reversibility of these coatings may be obtained by means of hydrogen plasma treatments generally utilised for etching purposes or by means of treatments in a plasma of oxygen pure or mixed with organic fluorinated compounds as CH_2F_2, CH_3F and CF_4.[15,16] An oxygen plasma is well known in plasma chemistry for its oxidising action on the exposed surfaces; in proper experimental conditions it is possible to convert the polymer functional groups in volatile compounds as CO_2, CO, SiF_4, without alteration of the metallic surface.

The protective effectiveness of the SiO_x films has been assessed on coated silver based alloy samples both by means of electrochemical impedance measurements and by means of the tarnishing test carried out in presence of H_2S vapours. Electrochemical impedance spectroscopy (EIS) is usually used to assess the protective properties and stability of organic films, because it may provide direct information on the corrosion rate.[17–19] A severe limitation of EIS technique, however, is the need of performing the measurement in an electrolyte that can affect the conservation state of an artefact of archaeological and historical value.

The visual monitoring of the samples submitted to the tarnishing test has been performed by means of an easy to use and cheap diagnostic tool based on digital photography and image processing. Other authors described examples of integrated digital photographic image acquisition, processing and analyses for cultural heritage research.[6,7] Specifically dealing with the monitoring of corrosion processes of metallic artefacts an automated remote monitoring photographic capture system has been developed within the European Union PROMET project, with the aim of monitoring protection systems efficiency for long time exposure of artefacts;[8] a measurement campaign is running at the Palace Armoury, Valletta, Malta on low carbon steels coupons protected with different methodologies.

Also the approach presented in this paper, although not able to provide quantitative measurement of the corrosion rate, is completely non-invasive and may be proposed for the *in situ* monitoring of restored real artefacts displayed in museum showcases. Moreover, the proposed imaging processing may help in evidencing the beginning of degradation phenomena, which require a quick restoration work.

Experimental

Plasma deposition of SiO_x protective films

A set of mirror polished discs ($20\times20\times2$ mm) of an Ag–Cu alloy (wt%: Ag 95.5, Cu 3.5) was coated by means of plasma enhanced chemical vapour deposition (PECVD) with SiO_x films.

The experimental apparatus used is a capacitively coupled parallel plate reactor. The samples were processed on the grounded electrode at the floating temperature reached by the plasma ($T<70^\circ C$), by applying the RF power, in the range of 50 to 250 W, to a gas mixture containing tetraethoxysilane (TEOS), oxygen and argon at different ratios (TEOS/O_2 ratios used 1 : 1, 1 : 10 and 1 : 20). The total flowrate was 41 sccm (standard cubic centimetres per min) and the pressure inside the process chamber was 5 Pa. Table 1 summarises the different conditions of deposition of the layers.

Film morphology and thickness were investigated by means of field emission scanning electron microscopy (FESEM); the film thickness ranges from 250 to 300 nm.

After the deposition of the SiO_x film, a small silver pad was deposited on the samples to evaluate the insulating properties. To this aim, the resistance and capacitance of the films were determined by means of a voltammeter that measures the magnitude and the phase of the sample impedance in the frequency range 0·01 Hz – 20 kHz.[20] The areas interested by the pad were obviously excluded by the image analyses.

The tarnishing test was performed by inserting the samples into a 5 dm^3 reaction chamber maintained at

Table 1 Ag–Cu sample identification numbers and experimental parameters used for deposition of SiO_x films

Sample number	TEOS/O_2 ratio, sccm	Input power, W
1	1:20	100
2	1:1	50
3	1:10	200
4	1:10	250
5	1:1	250
6	1:20	250
7	1:10	50
8	1:1	200
9	1:10	100
10	Uncoated	

20±2°C and containing a 10 ppm concentration of H_2S introduced with a single injection every day. The H_2S concentration in the chamber was measured by an electrochemical commercial sensor. The aging test lasted 35 days and the tarnishing evolution was checked every 7 days exposure periods.

Electrochemical impedance measurements

The protective effectiveness of the SiO_x thin films deposited in the different experimental conditions, was assessed by means of EIS. Impedance measurements were performed in an aerated 0·1M Na_2S solution at room temperature. Impedance spectra were recorded at the open circuit potential as a function of the exposure time, $1 h<t<168 h$, by applying a sinusoidal signal with 10 mV amplitude, in the frequency range 10^{-2}–10^5 Hz. Five points per decade were collected. An AgCl electrode as reference and a titanium counter electrode were used. The impedance spectra were analysed by means of a suitable equivalent electrical circuit model, as shown in Fig. 1, usually used to represent a metallic substrate coated with an organic layer[21,22] but with two constant phase elements (CPEs) which substitute the coating capacitance C_c (CPE 1), and the double layer capacitance C_{dl} (CPE 2), and which takes into account the phenomena related to the heterogeneous surface and diffusion processes. The impedance of a CPE is represented by the equation:

$$Z=C\times 1/(i\omega)^{\alpha}$$

where C is a constant depending on the specific analysed system, $i=(-1)^{1/2}$, ω is the angular frequency, α is a coefficient ranging between 0 and 1 and is related to the roughness of the surface. On smooth electrodes, $\alpha=1$ and $C=C_c$ or C_{dl}. R_{po} is the pore resistance due to electrolyte penetration in the damaged areas of the film, while the charge transfer resistance, R_{ct}, may be related to the defectiveness degree of coating.

The EIS data of coated metals are represented using Nyquist plots which depict the imaginary impedance, indicative of the capacitive and inductive character of the cell, versus the real impedance of the cell. Nyquist plots have the advantage that activation controlled processes with distinct time constants show up as unique impedance arcs and the shape of the curve provides insight into possible mechanism or governing phenomena. For comparison purpose, the EIS measurements were performed also onto the uncoated samples.

Tarnishing monitoring by contactless imaging tool

The tarnishing evolution was also assessed by using a completely non-invasive approach based on digital photography and image analysis.[23–25] The already

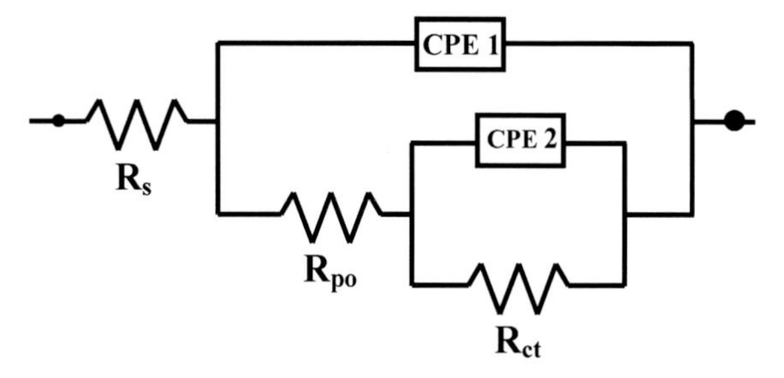

1 Equivalent circuit model of system metal/coating/electrolyte

proposed colour based image analyses require a controlled lighting to provide meaningful results.[17,18] Since the corrosion deriving from the protective layer failure is a localised one, a different approach sensitive to surface uniformity, rather than colour, may be advantageously applied. A 2D-FFT process may be used for highlighting the variations of the surface uniformity caused by corrosion phenomena and to extract a single parameter representative of the surface uniformity. The proposed parameter is computed by applying a four-step process.[26] Initially, each pixel of the image is processed to compute its brightness in the HSV colour space, thus obtaining a black and white image. Then a 2D-FFT transformation is applied to the picture of the surface obtaining another image where pixel luminance, computed as the modulus of the resulting 2D-FFT for that pixel, represents the energy content at different frequencies. The image centre represents the average value (i.e. a zero spatial frequency), while pixels with increasing x and y distance from the centre represent higher frequencies. Consequently, the surface image and the obtained FFT image for two samples, one undamaged and one corroded are significantly different. As a third step, the energy of pixels having the same x and y distances from the centre are added together to compute the total energy at the different frequencies. Eventually the cumulative energy is computed starting from the higher frequencies (i.e. starting from the FFT border). The damaged surfaces are associated to higher values of the energy at high frequencies, so that it is easy to select a predefined amount of energy (for example 5% in the test performed in this paper) and look for the position reached (for example 2% of distance from the centre for the undamaged sample and 25% for the damaged sample in the text illustrated below).

The FFT is a linear operator, consequently the result of the overall process is independent from the actual lighting value, provided that the camera sensor is linear too. Unfortunately this is not the case when using pictures generated by most of the commercial photographic cameras, which process the images with a gamma curve to compress the image dynamic range and obtain nice pictures. It is therefore important to employ a camera capable of producing RAW images, to use a processing program that does not apply gamma corrections and to correct the camera sensor non-linearities. In this study, the authors used a Panasonic Lumix G (Panasonic UK Ltd, Bracknell, UK), which produces images with 4000×3000 pixels and generates 12 bit/colour RAW images. The RAW images were developed by means of an open source program dcraw[27] generating uncompressed 16/colour TIFF images, processed in turn by means of a program specifically developed by the authors to perform the FFT processing and extract the final compact corrosion indicator. The sensor non-linearity was identified with the multiple exposure technique[25] and was corrected by the processing program.

All the results reported in this paper were obtained with the camera set on simple stand and by using a lighting obtained with a 4000 K fluorescent lamp. The lighting direction also plays an important role in determining the final result especially in the case of textured objects where an angled light may increase the contrast and therefore the energy dispersion also in the case of uncorroded samples. In all cases, a diffused light

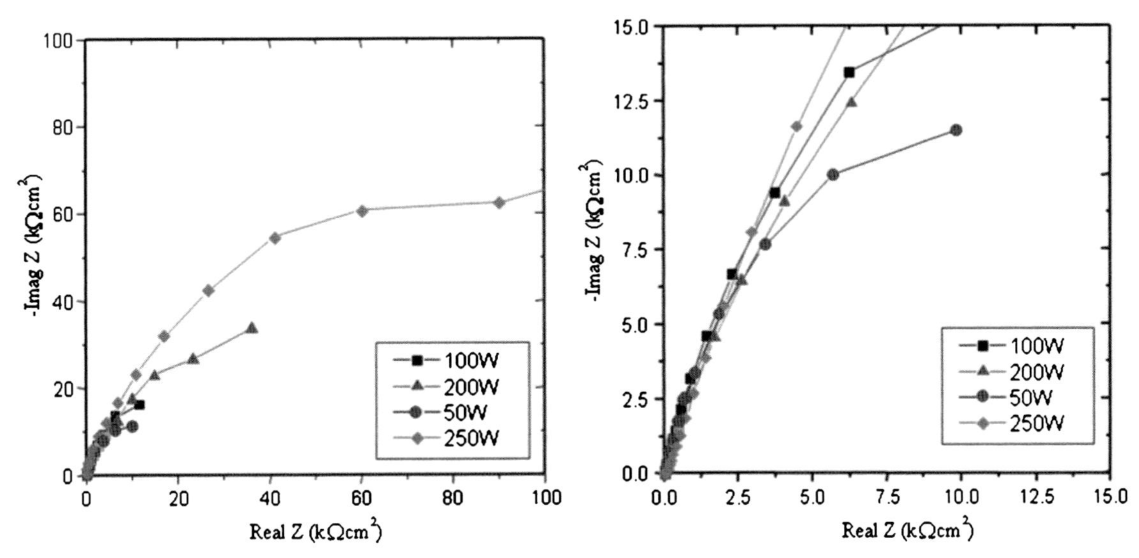

2 Nyquist plots recorded after 24 h immersion in 0·1M Na_2S solution on Ag–Cu samples coated in plasma fed with TEOS=1 sccm, O_2=10 sccm, Ar=29 sccm, at different input powers respectively from 50 to 250 W (samples 7, 9, 3, 4 of Table 1)

was used and the program equipped with the possibility of storing data and initial uniformity parameter to be used when judging the tarnishing evolution.

Results and discussion

Electrochemical evaluation of coatings protective effectiveness

Some results of the electrochemical characterisation in 0·1M Na_2S solution are shown in Figs. 2 and 3, for Ag–Cu samples SiO_x coated in different conditions.

Figure 2 shows the Nyquist plots after 24 h exposure to the aggressive environment for samples coated in plasma with the same TEOS/O_2 ratio, 1 : 10, but with different input powers.

In the case of SiO_x film deposited at low input power, sample 7, two capacitive loops are present: the first related to the electric properties of the coating and, the second to the Faradaic process at the substrate. In comparison to substrate, whose value of R_{ct} is $8{\cdot}50\times10^3$ Ω cm², the coated sample achieved a value of $7{\cdot}42\times10^4$ Ω cm², demonstrating improved corrosion resistance. A noticeable improvement in the protective effectiveness of the film was achieved by increasing the input power during deposition. As a matter of fact, the fitting of the data with the electrical equivalent circuit yields a higher R_{ct} value for sample 4, $4{\cdot}92\times10^5$ Ω cm², thus indicating better barrier properties of the coating.

In summary, the EIS characterisation of all the samples listed in Table 1 shows that the higher the discharge input power, the higher the barrier properties and the stability of the SiO_x thin films. Moreover, the lower the TEOS/O_2 ratio in the plasma reactor, the higher the protective effectiveness of the layers. This behaviour has to be related to an increasing inorganic nature of the deposited film, as found by the authors in previous papers,[28,29] by means of Fourier transform infrared spectroscopy characterisation of coatings that evidences how, by increasing the oxygen to polysiloxane ratio, an increase in the relative intensities of the peaks corresponding to Si–O–Si stretching and Si–O bending is observed.

Figure 3*a* and *b* shows the trend of R_{ct} as a function of immersion time for SiO_x films deposited on Ag–Cu samples in the different experimental conditions. The general decrease in R_{ct} with time indicates widening of

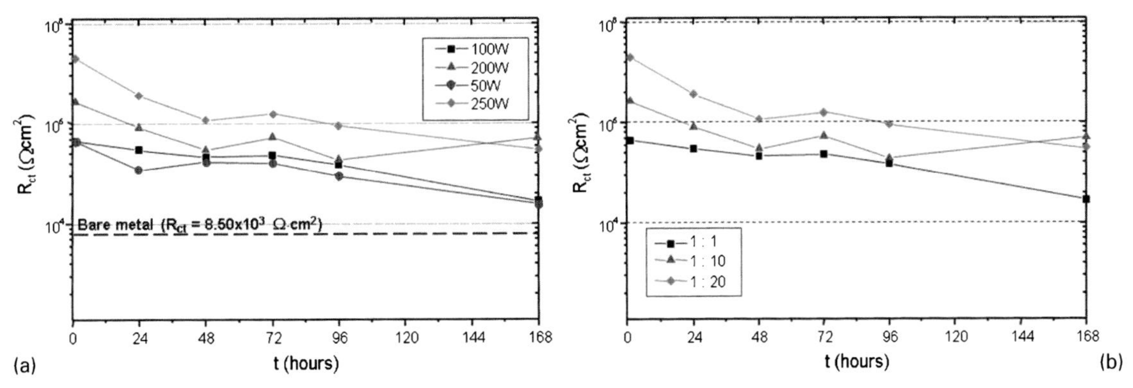

a in plasma fed with TEOS=1 sccm, O_2=10 sccm, Ar=29 sccm at different input powers, in the range 50–250 W; *b* in plasma fed with different TEOS/O_2 ratios at the same input power 50 W

3 R_{ct} trends recorded after increasing immersion periods in 0·1M Na_2S solution on Ag–Cu samples SiO_x coated under different experimental conditions

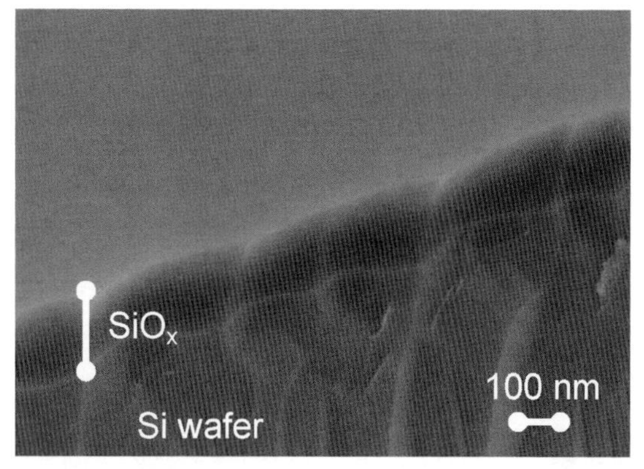

4 Image (FESEM) of cross-section of SiO$_x$ film deposited at 250 W in plasma fed with TEOS=1 sccm, O$_2$=20 sccm, Ar=29 sccm

the corroding area beneath the coating as a consequence of the progressive degradation of the coating itself.

As shown in Fig. 3*a*, the largest decrease in R_{ct} occurs in the first day of immersion for the samples deposited with the highest input power, then the situation stabilises and the R_{ct} values remain more than one order of magnitude higher than the value of the uncoated substrate also for immersion times of 168 h. All the films show a long-lasting protective effectiveness also if of lower entity in dependence on input power, as a matter of fact the lower values are obtained with sample 7.

The trends of the R_{ct} as a function of the immersion time for the SiO$_x$ films deposited at 50 W with increasing TEOS/O$_2$ ratios, as shown in Fig. 3*b*, evidence the beneficial effect of high oxygen content in the gas mixture on the barrier effects of the coatings also for long immersion times. The input power is, in any case, the parameter that mainly affects the barrier properties of the film.

Morphological analysis carried out by means of atomic force microscopy and scanning electron microscopy on uncorroded samples revealed that, if the deposition parameters are optimised, the deposited films cover uniformly the substrate surface and have a high adaptability degree to the surface morphology.[30] Figure 4 shows the FESEM image of a cross-section of a film deposited in plasma at 250 W with a TEOS/O$_2$ ratio of 1 : 20. The film that shows the best protective effectiveness among the ones tested in this study appears compact and well adherent with an amorphous structure.

Tarnishing evolution assessment and image processing results

The results of the tarnishing test, carried out in 10 ppm H$_2$S atmosphere, are in accordance with the electrochemical characterisation of the samples. The discharge input power is the parameter that mainly affects the barrier properties of the film against the aggressive vapours together with the oxygen content of the gas mixture.

Figure 5 shows the digital images of the samples listed in Table 1 after increasing exposure times, respectively 7, 14, 21, 28 and 35 days. A simple visual observation of the surfaces allows to state that: samples 6 and 1 coated at 250 and 100 W with a TEOS/O$_2$ ratio of 1 : 20 maintain almost unchanged the surfaces along all the exposure period; sample 4 coated at 250 W with a TEOS/O$_2$ ratio of 1 : 10 is subjected to negligible attacks; samples 3 and 9 coated respectively at 200 and 100 W with a TEOS/O$_2$ ratio of 1 : 10 show significant modifications of the exposed surfaces; sample 7 coated at 50 W with a TEOS/O$_2$ ratio of 1 : 10 is subjected to a noticeable tarnishing already after 14 days of immersion; the surfaces of samples 5 and 8 coated respectively at 250 and 200 W with a TEOS/O$_2$ ratio of 1 : 1 appear significantly corroded; the behaviour of sample 2 coated

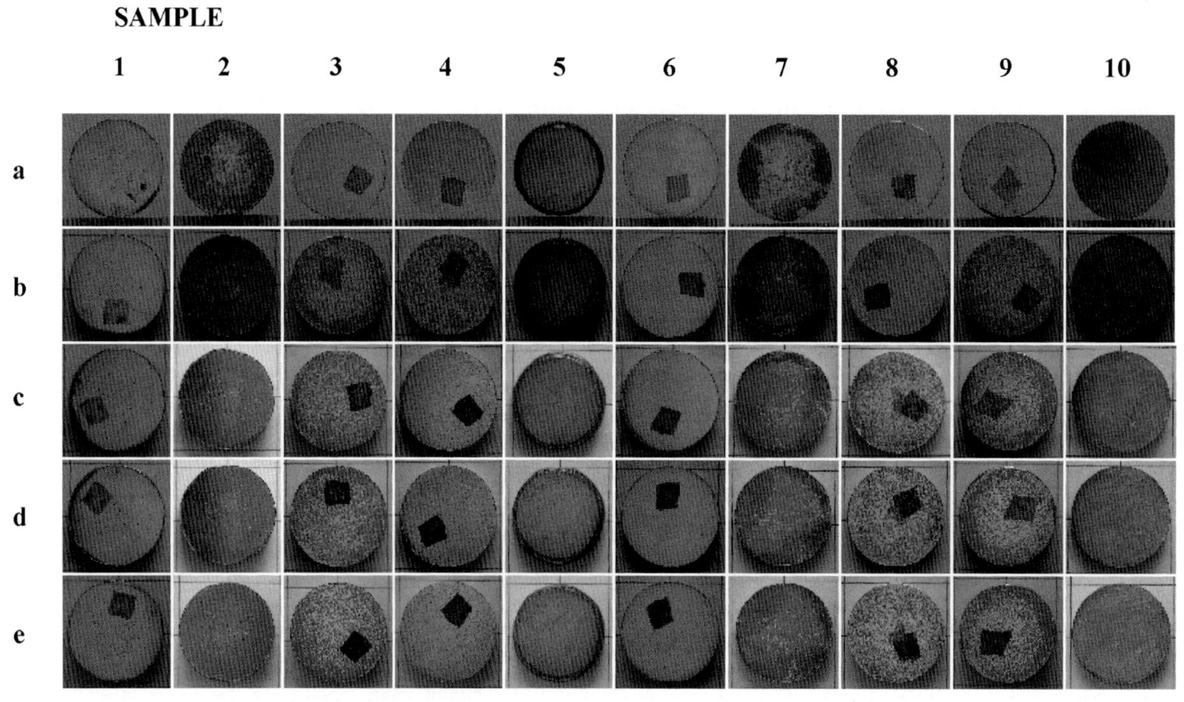

a 7 days; *b* 14 days; *c* 21 days; *d* 28 days; *e* 35 days

5 Digital images of Ag–Cu samples submitted to tarnishing test for increasing exposure times in 10 ppm H$_2$S atmosphere. See Table 1 for deposition conditions

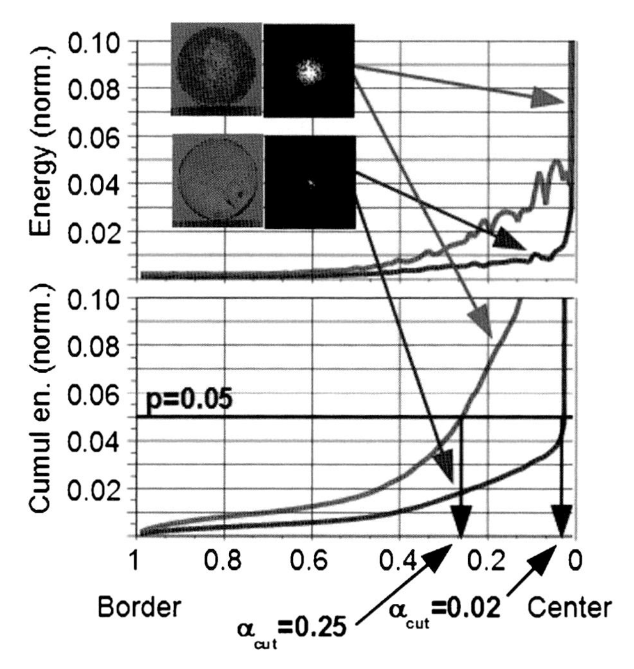

6 Digital images and elaborated FFT images for SiO_x coated Ag–Cu samples after 7 days exposure to 10 ppm H_2S atmosphere. Sample 2 (red lines) shows noteworthy tarnishing. Sample 1 (blue lines) shows negligible tarnishing. See Table 1 for deposition conditions

at 50 W with a TEOS/O_2 ratio of 1 : 1 is very similar to the one of sample 10, the uncoated Ag–Cu alloy, an increasing tarnishing of the surface takes place from the first 7 days exposure.

To follow the tarnishing evolution, highlighting the variations of the surface uniformity caused by corrosion phenomena, the above illustrated non-invasive approach based on digital photography and image analysis has been applied to the set of digital images of Fig. 5. The 2D-FFT process allows to extract a single parameter representative of the surface uniformity that may be related, in the case of coated samples, to the coating layer stability.

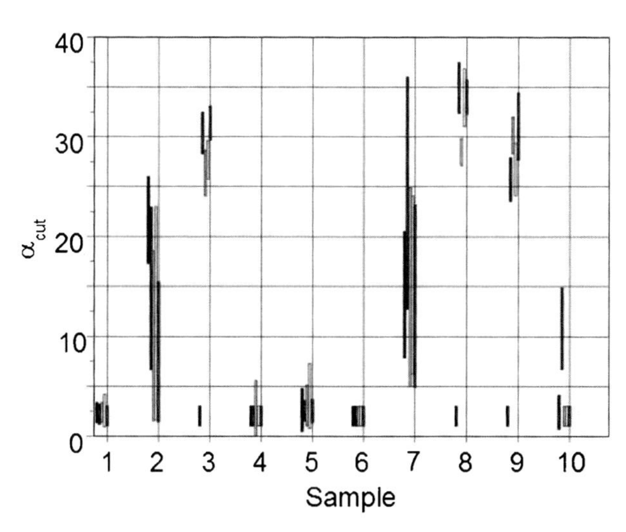

7 Results of image processing obtained on digital images of Fig. 5 of Ag–Cu samples submitted to tarnishing test for increasing exposure times to 10 ppm H_2S atmosphere: 7 days (black bar), 14 days (blue bar), 21 days (green bar), 28 days (yellow bar) and 35 days (red bar). See Table 1 for deposition conditions

As an example of the image processing application, Fig. 6 shows the digital images and FFT images of two SiO_x coated Ag–Cu samples after 7 days exposure to 10 ppm H_2S atmosphere: sample 1 showing negligible tarnishing and sample 2 showing a noteworthy extension of the tarnished areas.

For each image, the processing was performed by analysing different surface portions with dimension 256×256 pixels corresponding to an area of $3{\cdot}5\times3{\cdot}5$ mm and computing mean and standard deviation of all the values obtained on the same sample. The corroded surfaces are associated to higher values of the energy at high frequencies, so that it is easy to select a predefined amount of energy, 5%, and look for the position reached, 2% of distance from the centre for the undamaged sample and 25% for the damaged sample. A value of energy corresponding to 99·2% has been used to compute the α_{cut} value.

Figure 7 summarises the results of the image processing applied to the digital images of the surface of the 10 samples shown in Fig. 5.

The behaviour of each sample is represented by bars corresponding to the α_{cut} average value plus/minus standard deviation. Higher α_{cut} values reflect a noteworthy lack of homogeneity on the surfaces, for example an increase in surface roughness due to localised corrosion phenomena.

Samples 7 and 2 coated with the films with the lower barrier properties show the highest values of the bars in Fig. 7, thus indicating a continuous increase in the corroded area after the coating failure. On the contrary, sample 6, coated with the SiO_x film with the best protective effectiveness, deposited at high input power and with a high oxygen content in the gas mixture, shows no variation in the bar values, thus indicating a complete stability of the coated surface.

Intermediate behaviours are observed for the other samples, thus indicating for samples 3, 8 and 9 a tendency to progressive tarnishing after 14 days of exposure, while samples 1, 4 and 5 show a higher uniformity of the surface and consequently a lower degradation of the coating.

The utilised approach, essentially sensitive to inhomogeneous surfaces does not reveal high variations in the values of the uncoated sample 10, which darkens to a great extent from the beginning. Anyway in all the other cases the proposed diagnostic tool may be efficiently employed for pointing out in a simple and non-destructive way the onset of corrosion phenomena.

Conclusions

SiO_x coatings deposited by PECVD may be proposed for the protection of silver artifacts from tarnishing, because of their good barrier properties, good aesthetic appearance and reversibility.

Electrochemical impedance spectroscopy has been used to assess the protective effectiveness of the film in a quantitative way. Moreover, the imaging based qualitative approach proposed for the tarnishing evolution assessment proved to be an effective diagnostic tool to evaluate the stability of the films and onset of localised corrosion phenomena that modify the surface roughness.

SiO_x coatings have been deposited starting from tetraethoxysilane as a precursor in different deposition conditions. The experimental findings show that the higher the input power applied in the PECVD reactor, the higher the percentage of oxygen in the gas mixture, the better the protective effectiveness against sulphur containing environments of the obtained layers.

References

1. C. Degrigny, M. Jerôme and N. Lacoudre: *Corros. Australas*, 1993, **2**, 16–18.
2. H. A. Ankersmit, G. Noble, L. Ridge, D. Stirling, N. H. Tenne and S. Watt: in 'Tradition and innovation: advances in conservation', (ed. R. AshokSmith *et al.*), 7–13; 2000, London, IIC.
3. V. Costa: *Rev. Conserv.*, 2001, **2**, 18–34.
4. A. Lins and N. McMahon: 'Current problems in conservation of metal antiquities', 135–162; 1989, Tokyo, Tokyo National Institute of Cultural Properties.
5. C. Degrigny: *J Solid State Electrochem.*, 2010, **14**, 353–361.
6. C. Degrigny and D. Witschard: in 'Medieval reliquiary shrines and precious metal objects', (ed. K. Anheuser *et al.*), 9–16; 2000, London, Archetype.
7. ICOM-CC: 'Terminology to characterize the conservation of tangible cultural heritage', available at: http://www.icom-cc.org/54/document/terminology-to-characterize-the-conservation-of-tangible-culturalheritage-english/?id=368 (accessed 18 April 2009).
8. M. Gilberg: *IIC-CG Newsletter*, 1987, 12–3.
9. T. Stambolov and E. Moll: ICOM Committee for Conservation Plenary Meeting, ICOM Report 69/37, Amsterdam, The Netherlands, 1969, 915–919.
10. A. Domenech-Carbò: *J. Solid State Electrochem.*, 2010, **14**, 349–351.
11. M. C. Bernard, E. Dauvergne, M. Evesque M. Keddam and H. Takenouti: *Corros. Sci.*, 2005, **47**, 663–679.
12. E. Cano, J. Barrio, D. M. Bastidas, S. Fajardo and J. M. Bastidas: in 'MetalEspaña '08. Congreso de conservación y restauración del patrimonio metálico', (ed. J. Barrio *et al.*), 74–79; 2009, Madrid, UAMCSIC.
13. F. Fracassi, R. d'Agostino, F. Palumbo, E. Angelini, S. Grassini and F. Rosalbino: *Surf. Coat. Technol.*, 2003, **174–175**, 107–111.
14. R. d'Agostino, F. Fracassi, F. Palumbo, E. Angelini, S. Grassini and F. Rosalbino: *Plasma Process. Polym.*, 2005, **2**, 91–96.
15. S. Grassini, E. Angelini, R. d'Agostino, F. Palumbo and G. M. Ingo: in 'Strategies for saving our cultural heritage', Proc. Int. Conf. on 'Conservation strategies for saving indoor metallic collections', Cairo, Egypt, February–March 2007, Applied Laser Spectroscopy group, Vol. 1, 127–131.
16. F. Fracassi, R. d'Agostino, A. Fornellli, F. Illuzzi and T. Shirafuji: *J. Vac. Sci. Technol.*, 2003, **21**, (3), 638–642.
17. E. Cano, D. Lafuente and D. M. Bastidas: *J. Solid State Electrochem.*, 2010, **14**, 381–391.
18. F. Mansfeld: *J. Appl. Electrochem.*, 1995, **25**, 187–196.
19. J. N. Murray: *Prog. Org. Coat.*, 1997, **30**, 225–230.
20. A. Carullo, S. Grassini and M. Parvis: *IEEE Trans. Instrum. Meas.*, 2009, **58**, 120–130.
21. F. Mansfeld, M. W. Kendig and S. Tsai: *Corrosion*, 1982, **38**, 478–485.
22. J. R. Macdonald: 'Impedance spectroscopy'; 1987, New York, John Wiley & Sons.
23. J. Winter, C. Maines and J. Dickerson: in 'Imaging the past – electronic imaging and computer graphics in museums and archaeology – British Museum Occasional Paper 114' (ed. T. Higgins *et al.*); 1996, London, British Museum Press.
24. C. Manders and S. Mann: Proc. 8th IEEE Int. Symp. on 'Multimedia', pp. 1–10, San Diego, CA, USA, December 2006, IEEE.
25. M. D. Grossberg and S. K. Nayar: *IEEE Trans. Pattern Anal. Machine Intell.*, 2003, **25**, (11), 1455–1467
26. E. Angelini, S. Grassini, D. Mombello, A. Neri and M. Parvis: *Appl. Phys. A*, 2010, DOI: 10.1007/s00339-010-5669-1.
27. Dcraw program for image RAW conversion, available at: http://www.cybercom.net/dcoffin/dcraw/.
28. E. Angelini, F. Fracassi, R. d'Agostino, S. Grassini and F. Rosalbino: in 'Trends in electrochemistry and corrosion at the beginning of the 21th century', (ed. J. M. Costa *et al.*), Vol. 1, 979–999; 2004, Madrid, Universitat de Barcelona.
29. C. Voulgaris, E. Amanatides, D. Mataras, S. Grassini, E. Angelini and F. Rosalbino: *Surf. Coat. Technol.*, 2006, **200**, 6618–6622.
30. E. Angelini, S. Grassini, F. Rosalbino, F. Fracassi, S. Laera and F. Palumbo: *Surf. Interf. Anal.*, 2006, **38**, 248–251.

On unexpected colour of lead sculptures in Queluz: degradation of lead white

M.-C. Bernard[1], V. Costa[2] and S. Joiret*[1]

Although the usual aspect of lead surfaces is light grey, in the gardens of the 'Palácio Nacional de Queluz', Portugal, some of the sculptures present an atypical red brown appearance. In order to investigate this phenomenon, lead coupons previously prepared with different surface finishes have been exposed in nearby statues presenting both surface aspects. After 15 months, visible surface transformations have been noticed for certain locations, and Raman microscopy has been used to characterise the surface products. Lead(II) oxide near the metal and salt layer at the outer part has been found for the white patina. For the red patina, a mixture of lead(IV) oxide and an uncommon lead oxycarbonate, probably shannonite ($PbCO_3.PbO$), has been found. Complementary *in situ* Raman analyses performed during electrochemical oxidation of lead, lead covered with hydrocerussite and pure hydrocerussite have shown that the presence of both lead and hydrocerussite is required to reproduce the field experiment.

Keywords: Plattnerite, Shannonite, Electrochemical oxidation, *In situ* Raman spectroscopy, Cavity microelectrode

This paper is part of a special issue on corrosion of archaeological and heritage artefacts

Introduction

Queluz Palace is often called the Versailles of Portugal. One of the attractions of this palace is its beautiful decorated gardens. The layout was designed by Jean Baptiste Robillion, who chose a semiformal Italian style and decorated with a boating canal and numerous fountains, and period reproduction sculptures, which were made of unalloyed lead by John Cheere. They were designed in the mid-eighteenth century. Lead exposed to atmosphere is expected to take a white patination layer of lead(II) oxide and lead hydroxycarbonate. However, some of the statues present a red brown colour.

This work is an attempt to understand this phenomenon through the analysis of coupons exposed for 15 months in different locations in the Palace.

Experimental

Sets containing lead coupons (Weber Metaux, Paris, France), previously treated to obtain different surface finishes – mechanically polished with SiC paper in water; coated with patination oil (which are disregarded in this study); set in 1M acetic acid atmosphere for 24 h; set in 1M NaOH atmosphere for 24 h and prepared by immersion in 0·1M H_2SO_4 solution and rinsing with water – have been exposed for 15 months in several places, in nearby sculptures presenting both types of surface coloration (red brown and grey) as well as inside the Palace.

Lead (Goodfellow, Lille, France), used for the electrochemical measurements, was first cleaned with hydrochloric acid, carefully washed, polished on SiC paper (1200 grade) with water and dried in argon flux.

A classical three-electrode cell is used, with a platinum counter electrode and a mercurous sulphate reference electrode [V(MSE)=+0·650 mV(SHE)]. Potentials are applied with an Eco Chemie Autolab 30 potentiostat (Eco Chemie, Utrecht, The Netherlands). The electrolyte is a pH 8·4 borate buffer solution. Electrochemical hydrocerussite oxidation was performed in a cavity microelectrode, allowing polarisation of a small amount of powder.[1]

The Raman spectrometer equipment is a HORIBA Jobin-Yvon LabRam (Longjumeau, France) consisting of an Olympus BX40 microscope confocally coupled to a 300 mm focal length spectrograph. The spectra were obtained with 632 817 nm radiation from an internal 10 mW HeNe laser with neutral density filters, 0·7 mW remaining at the sample, to avoid any thermal effect. For *in situ* studies, a ×50 ULWD objective allows recording Raman spectra with a working distance of 8 mm. The electrode is then covered by a 3 mm thick borate solution layer.

Results and discussion

Visual examination

Figure 1 shows the aspect of the lead coupons before and after 15 months' exposure. Although the overall appearance of the coupons exposed indoor (Fig. 1B) is basically the same as in their initial state, some evident changes can be observed in those kept outdoors. In the case of sets exposed near light grey sculptures, the lead

[1]Laboratoire Interfaces et Systèmes Electrochimiques, UPR 15 du CNRS, Université P. et M. Curie, case 133, 4 Place Jussieu, 75252 Paris, France
[2]Conservare, 21 rue des Cordeliers, 60200 Compiègne, France

*Corresponding author, email suzanne.joiret@upmc.fr

Published by Maney on behalf of the Institute
Received 15 December 2009; accepted 2 May 2010
DOI 10.1179/147842210X12732285051276

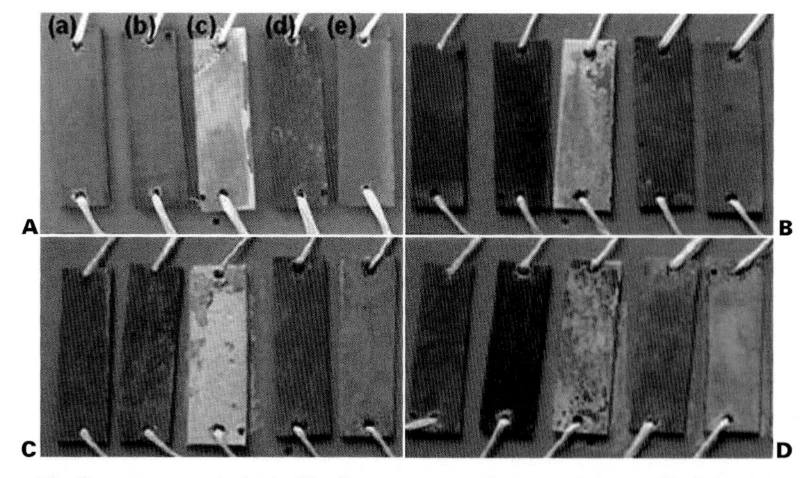

(A) before exposure; (B) after exposure, indoor; (C) after exposure, close to statues with light grey surface; (D) after exposure, close to statues with red brown surface

1 Aspect of lead coupons before and after 15 months' exposure [aspect of different surface pretreatments in (A): *a* mechanically polished, *b* coated with patination oil, *c* prepared in acetic acid atmosphere, *d* prepared in NaOH atmosphere and *e* prepared by immersion in H_2SO_4]

surfaces present a slight increase in the amount of white product (Fig. 1C). Concerning the sets maintained close to red brown statues (Fig. 1D), the most remarkable change has been observed for the coupons previously treated in acetic and sulphuric acid atmospheres (third and fourth from left).

Ex situ micro-Raman analysis

Samples from the three abovementioned locations have been analysed, except the coupons covered with patination oil, which presented fluorescence and are not considered in this study. Raman analyses have been performed on the exposed surface and on the cross-section. Results are summarised in Table 1.

The cross-section analysis shows the stratigraphic distribution of the surface products. Close to the metal, all samples present a thin layer of litharge. In a few cases, the presence of massicot has also been detected. This oxide layer is thicker for samples pretreated in sodium hydroxide atmosphere, giving the orange colour of coupon d (Fig. 1A).

At the outer part, a salt layer is found, the composition of which depends on pretreatment and exposure location. For coupon e (Fig. 1A), pretreated in sulphuric acid, pure anglesite $PbSO_4$ is analysed in the indoor coupon, and a mixture of anglesite and hydrocerussite in various proportions is found in outdoor coupons. For coupon d (Fig. 1A) pretreated in sodium hydroxide atmosphere, the growth of a basic carbonate layer is attested, even in indoor exposure, which indicates that the litharge layer close to the metal seems not to be passivating.

For coupon c (Fig. 1A) pretreated in acetic acid atmosphere, the growth of a thick hydroxycarbonate layer is the consequence of exposure to a carboxylic acid, as already mentioned in the literature.[1] The nature of this hydroxycarbonate depends on the exposure: plumbonacrite for indoor and hydrocerussite for outdoor. This autocatalytic process leads to a thick carbonate layer in a short period of time.[5] Moreover, the red stains located at an intermediary part of the patina layer, always in the presence of a thick (20–50 μm) hydrocerussite layer, give the spectrum named X in Table 1 and displayed in Fig. 2. Three distinct features appear on this spectrum: two thin, intense bands respectively situated at 160 and 1055 cm^{-1} and a large unresolved band extending from 450 to 550 cm^{-1}. The two first bands can be attributed to a carbonate

Table 1 Compounds detected by Raman analysis for different pretreatments and exposure sites

	Exposure site		
Pretreatment	Indoor	Outdoor, close to light grey statue	Outdoor, close to red brown statue
Mechanically polished	Litharge[2] Carbonates	Litharge[2] Massicot[2] Hydrocerussite[3]	Litharge[2] Carbonates
Exposure in acetic acid atmosphere	Litharge[2] Plumbonacrite[3]	Litharge[2] Massicot[2] Hydrocerussite[3] $CaCO_3$	Litharge[2] Massicot[2] Hydrocerussite[3] Spectrum X
Exposure in sodium hydroxide atmosphere	Litharge[2] $2PbCO_3.NaOH$[3]	Litharge[2] Carbonates	Litharge[2] Hydrocerussite[3]
Immersion in sulphuric acid	Litharge[2] Anglesite[4]	Litharge[2] Anglesite[4] Hydrocerussite[3]	Litharge[2] Anglesite[4] Hydrocerussite[3] Spectrum X

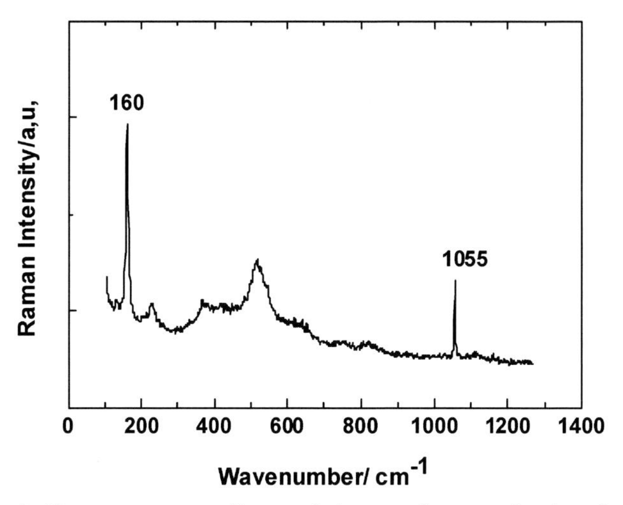

2 Raman spectrum X recorded on red spots developed on lead coupons previously treated in acidic solutions and exposed near statues with red brown aspect

anion, while the third feature has already been reported in Ref. 2 as the Raman signature of lead(IV) oxide.

In situ Raman analysis during electrochemical oxidation

In order to understand the relationship between spectrum X and lead oxidation, electrochemical oxidation has been applied to three cases: pure metal, metal covered with hydrocerussite and pure hydrocerussite.

In the case of pure lead, the sample has been exposed to borate buffer solution and anodically polarised at fixed potential values. The Raman spectra recorded at those potential are given in Fig. 3. At open circuit potential (OCP) value [−0·8 V(MSE), Fig. 3*a*], pure lead is in contact with solution, and no features are observed. For electrochemical oxidation at −0·3 V(MSE) (Fig. 3*b*), formation of litharge (147 and 345 cm^{-1} bands) is observed. For more positive potential values, −0·1 V(MSE) (Fig. 3*c*), bands corresponding to litharge begin to decrease, while a bump at 400–550 cm^{-1} starts to grow. Anodic oxidation of lead into lead(IV) oxide should take place in this potential range.

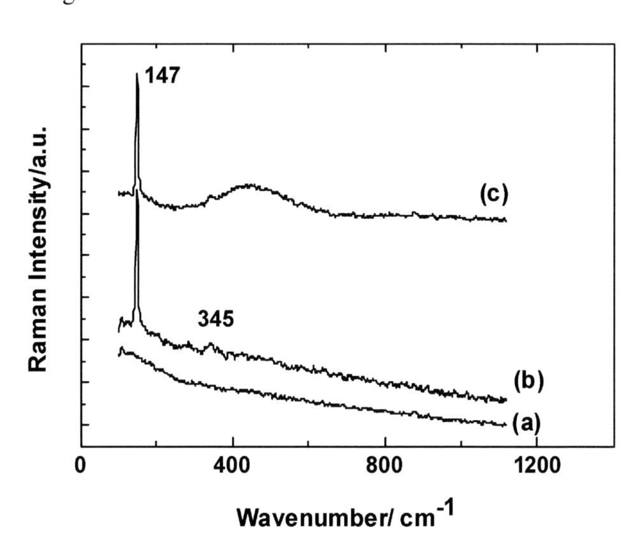

3 *In situ* Raman spectra recorded for pure lead in borate buffer solution, polarised at *a* −0·8 V(MSE), *b* −0·3 V(MSE) and *c* −0·1 V(MSE)

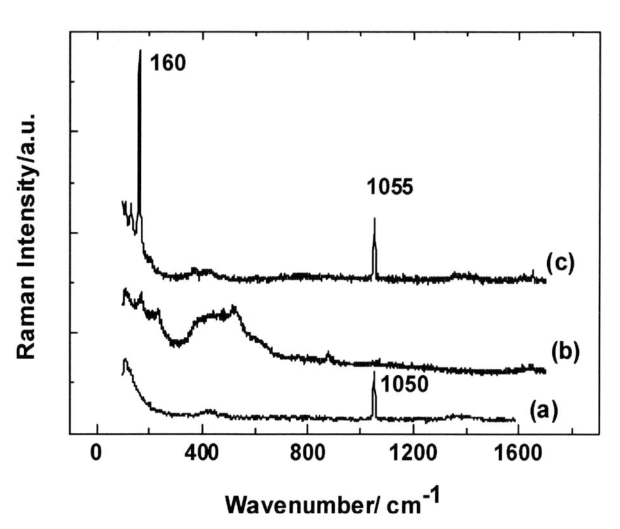

4 *In situ* Raman spectra recorded *a* at OCP of −0·9 V(MSE) (blue) and at +0·7 V(MSE) at two different locations: *b* red brown product and *c* transparent crystals

In situ Raman spectra have also been recorded on a lead coupon covered with a thick hydrocerussite layer, previously grown in the authors' laboratory, as explained and characterised in Ref. 1. They are displayed in Fig. 4. At OCP [−0·9 V(MSE)], the recorded spectrum (Fig. 4*a*) is one of pure, white hydrocerussite. After oxidation at +0·7 V(MSE), two new phases appear in the optical microscope used to focus the laser light: one consists of transparent crystals (Fig. 4*c*), and the other corresponds to a brown product (Fig. 4*b*).

The Raman spectrum in Fig. 4*c* recorded on isolated crystals is to be related to carbonate formation through the band at 1055 cm^{-1}, different from that on hydrocerussite located at 1050 cm^{-1}; however, this spectrum is not reported in the literature.

Those transparent crystals are not stable when polarisation is turned off or the electrode is taken out of the solution; within 1 h, crystals disappeared from the surface. The presence of an intense band close to the one of lead(II) oxide (160 cm^{-1} for carbonate and 147 cm^{-1} for PbO) leads us to propose the formation of shannonite[6] $PbCO_3$.PbO. Those types of oxycarbonates are known as transient species during basic lead carbonate thermal decomposition.[7]

The third spectrum (Fig. 4*b*) characterising the brown compound giving the colour of the sample is formed by two large bands centred at 420 and 550 cm^{-1} respectively. This Raman spectrum is the one of PbO_2 oxide in nanometric dimensions and substoichiometric. This type of Raman spectrum modification, in function of particle size and stoechiometry, has been reported for SnO_2.[8] This Raman spectrum has also been recorded by Aze *et al.*[9] when studying the degradation of lead pigments. The poor resolution of this spectrum does not allow its attribution to a crystalline form of PbO_2 like plattnerite or scrutynite.

Nevertheless, oxidation of lead covered with hydrocerussite leads to the formation of the two species, carbonate and oxide, already identified but mixed together in the spectrum in Fig. 2.

Furthermore, scanning electron microscope examinations of the electrode after polarisation and of the

5 Images (SEM) of anodically oxidised lead + hydrocerussite (left) and coupon from central fountain (right)

reddish coupon from central fountain are given in Fig. 5.

The laboratory sample shows that lead(IV) oxide keeps the shape of former hydrocerussite crystals, while the coupon presents a rather unusual aspect for a corrosion layer. Moreover, X-ray dispersive analysis has been performed to evidence an element likely to play the oxidative role, but only traces of Ca, Si and Al have been recorded.

Electrochemical oxidation of pure hydrocerussite powder

Pure hydrocerussite has been submitted to oxidation in the same borate solution. At 0·7 V(MSE), the white powder inside the cavity microelectrode turns brown and gives the spectrum in Fig. 4*b*, a characteristic of lead(IV) oxide. The potential needed to record this spectrum is the same as that of the oxidation of lead + hydrocerussite, higher than that of pure lead. However, no traces of carbonate, except the one from hydrocerussite before oxidation, have been recorded.

It can then be proposed that the red colour of the white patina is due to oxidation of hydrocerussite by a strong oxidant, like the already reported process in chlorinated waters.[10] Patina oxidation gives the brown compound (PbO_2), but due to the presence of lead underneath, this oxidation process gives also an oxycarbonate, shannonite, which is not formed when hydrocerussite alone is submitted to the oxidation process.

Conclusions

The unusual red brown colour observed on the surface of some lead statues in the gardens from Queluz palace can be attributed to the presence of lead(IV) oxide. This compound, which develops only under action of a strong oxidant agent, has also been detected on lead coupons preliminarily treated in acidic atmosphere and exposed for 15 months at the same locations, attesting for the role played by the environmental conditions. The presence of an oxycarbonate in the patina layer has to be related to the underneath metal oxidation, as this compound has only be identified on laboratory samples when both metal and hydrocerussite were present together.

Acknowledgements

This project was made possible by World Monuments Fund Britain (London, UK) and the Paul Mellon Estate (Middlleburg, VA, USA).

References

1. M. C. Bernard, V. Costa and S. Joiret: *e-Preserv. Sci.*, 2009, **6**, 101–106.
2. L. Burgio, R. J. H Clark and S. Firth: *Analyst*, 2001, **126**, 222–227.
3. M. H. Brooker, S. Sunder, P. Taylor and V. J. Lopata: *Can. J. Chem.*, 1983, **61**, 494–502.
4. R. L. Frost, J. T. Kloprogge and P. A. Williams: *Neues Jahrb. Miner. Mon.*, 2003, **12**, 529–542.
5. G. C. Allen and L. Black: *Br. Corros. J.*, 2000, **35**, (1), 39–42.
6. A. C. Roberts, J. A. R. Stirling, G. J. C. Carpenter, A. J. Criddle, G. C. Jones, T. C. Birkett and W. D. Birch: *Mineral. Mag.*, 1995, **59**, 305–310.
7. D. A. Ciomartan, R. J. H. Clark, L. J. McDonald and M. Odlyha: *J. Chem. Soc. Dalton Trans.*, 1996, **18**, 3639–3645.
8. L. Abello, B. Bochu, A. Gaskov, S. Koudryavtseva, G. Lucazeau and M. Roumyantseva: *J. Solid State Chem.*, 1998, **135**, (1), 78–85.
9. S. Aze, J. M. Vallet, M. Pomey, A. Baronnet and O. Grauby: *Eur. J. Mineral.*, 2007, **19**, 883–890.
10. H. Liu, G. V. Korshin and J. F. Ferguson: *Environ. Sci. Technol.*, 2008, **42**, 3241–3247.

Long term corrosion of aluminium materials of heritage: analysis and diagnosis of aeronautic collection

E. Rocca*[1], F. Mirambet[2,3] and C. Tilatti[4]

The large use of aluminium alloys for the twentieth century induces that numerous components and objects in aluminium are registered in many collections in museums or are classified as historic buildings. In collaboration with the Air and Space Museum at Le Bourget (France), the objective of the authors' work was to identify and study the different processes of corrosion of Al alloys occurring in the aircrafts collection from 1930 to now. Chemical and metallographic analyses allowed the identification of various alloys used in different aircrafts. The long-term corrosion products are mainly constituted by amorphous $Al(OH)_3$. The corrosion behaviour of the complex systems 'alloys/corrosion layer' was evaluated by electrochemical measurements. The results show that thick corrosion layers have a poor influence on the corrosion rate of the metallic substrate. On the other hand, the presence of painting residues with corrosion products is beneficial for the conservation because of the presence of mineral inhibiting compounds.

Keywords: Aluminium, Corrosion, Artefacts, Cultural heritage, Electrochemistry

This paper is part of a special issue on corrosion of archaeological and heritage artefacts

Introduction

With 25 million tons produced per year, Al is the second used metal in world after Fe in steels. Its use rapidly grew after the beginning of the twentieth century in the transport industry because of its low density. Thus, for more than 50 years, many objects or elements are entered in the museum collection or are listed as historic buildings.

Because these materials are considered as modern materials, generally unknown in many heritage collections, only few studies are devoted to the problems of their conservation. Consequently, the curators and restorers are usually resourceless when operations of conservation and restoration should be undertaken on the Al objects. It is the reason for which curators in charge for Al collections ask laboratories for solutions to treat the problems of Al artefacts.

The Air and Space Museum of Le Bourget (France) near Paris has one of the more important aircraft collections all over aviation history, from pioneers to nowadays. In fact, light Al alloys with good mechanical properties were used very early, in the 1930s, in aircrafts. Unfortunately, this kind of alloys is very sensible to corrosion phenomenon. For these 'aeronautic' alloys, the corrosion behaviour is closely related to the presence of particles, also named intermetallic phases, inside the Al matrix.[1] The very small intermetallic phases, with a nanometric size, are necessary to provide the mechanical resistance, but the large phases with a micrometric size are responsible for the drastic decrease in the corrosion resistance of the metallic material. Consequently, important deteriorations can be observed on many aircrafts after several 10 years of exposition of storage.[2]

On the other hand, the important size of aircrafts needs to expose or to store them outside or in sheltered conditions, but without control of environmental conditions, which constitutes a worse factor (Fig. 1).

In the framework of the French National Research Program on materials of cultural heritage, the purpose of this work was to analyse and study different kinds of 'aeronautic' Al artefacts. The goal was to propose a first methodology to diagnose the conservation state of these materials after long term corrosion.

Representative samples of different period were taken on some aircrafts of the collection belonging to the Air and Space Museum.

In a first stage, the work was devoted to the metallographic and chemical analysis of different 'aluminium alloys/corrosion layer and surface treatment' systems sampled on the aircrafts. From 1930 to 1970, eight aircrafts were selected with several types of corrosion morphologies to perform sampling. In this article, the authors present some representative cases of this study.

Then, the authors have characterised the chemical reactivity, especially the corrosion rate of the metal with or without corrosion products, in corrosive immersed conditions by electrochemical measurements.

[1]Corrosion Team, Institut Jean Lamour UMR 7198, BP70239, 54506 Vandoeuvre-Les-Nancy, France
[2]LRMH, 29 rue de Paris, 77420 Champs-sur-Marne, France
[3]C2RMF UMR CNRS 171, 6 rue des Pyramides, 75001 Paris, France
[4]Musée de l'Air et de l'Espace, Aéroport du Bourget, BP 173, 93352 Le Bourget Cedex, France

*Corresponding author, email emmanuel.rocca@lcsm.uhp-nancy.fr

Published by Maney on behalf of the Institute
Received 19 January 2010; accepted 1 April 2010
DOI 10.1179/147842210X12710800383765

1 Lockheed Super Constellation (Second World War) stored outside

Analysis of corrosion morphology of sample collection of Al artefacts

The experimental conditions of metallographic preparation and analysis by scanning electron microscopy (SEM) coupled with electron dispersive X-ray spectroscopy (EDX) and electron probe microanalysis (EPMA) are detailed in previous studies.[3]

A first kind of degradation phenomenon is lamellar corrosion and is generally characterised by a very thick layer of corrosion products. Figure 2 presents the macroscopic and microscopic aspect of sample taken from a spar of the Vautour aircraft (1956). As can be seen in Fig. 2*b*, the visual aspect after long term corrosion is clearly foliated and lamellar. The metallographic observations displayed in Fig. 2*c* show that the growth of the corrosion products at the grain boundaries leads to the destruction of the metallic piece.[4,5] The EPMA analysis confirms that the metal is an Al–Cu alloy with ~3·75 wt-% and also contains some traces of 0·1 wt-%Fe and 0·6 wt-%Mn. The chemical composition of the intermetallic precipitates is difficult to analyse with accuracy by EDX because of their small size. Nevertheless, two kinds of particles can be detected: a first with a composition close to Al_2Cu and a quaternary precipitate Al_3(Fe,Cu,Mn). At the surface of the corrosion layer, some small phases containing Cr, Zn and Ti are some residues of paint (Fig. 2*d*). Cr and Zn are classical elements used in mineral corrosion inhibitor in paints, and Ti comes from Ti oxide particles used as white pigments in paints.

Another kind of corrosion morphology is observed on samples of 'Point d'Interrogation' aircraft, which is a Breguet 19 (1929). In the macroscopic photo displayed in Fig. 3*a* and *b*, this sample only seems to present some corrosion spots. Nevertheless, the metallographic analysis of the cross-section performed in several locations reveals important internal corrosion at the grain boundaries of the alloy. Figure 3*d* shows that the half of the original thickness of the metal is affected by internal oxidation. On the surface, the homogeneous presence of S, detected by EDX, proves that this piece has been anodised in sulphuric acid that is the classical anticorrosion treatment of Al alloys.[6] The EPMA analyses of metal reveal that the alloy has approximately the same composition than that of Vautour (3·8 wt-%Cu and 0·38 wt-%Mn). It is important to note that this kind of intergranular corrosion is as destructive as the lamellar corrosion observed in the Vautour samples, but the piece has preserved its original size as a whole.

The third analysed sample is a hatch located under the wing of the Mirage IV aircraft (1965). The morphology of the lamellar corrosion is clearly observed on the outside face of the hatch (Fig. 4*a*). On the contrary, the inside face displayed in Fig. 4*b* is much less corroded. In fact, the metallographic analysis of this face reveals the presence of a homogeneous oxide layer containing S, which is an anodising pretreatment before applying the paints. Two kinds of mineral residues of paints can be distinguished:

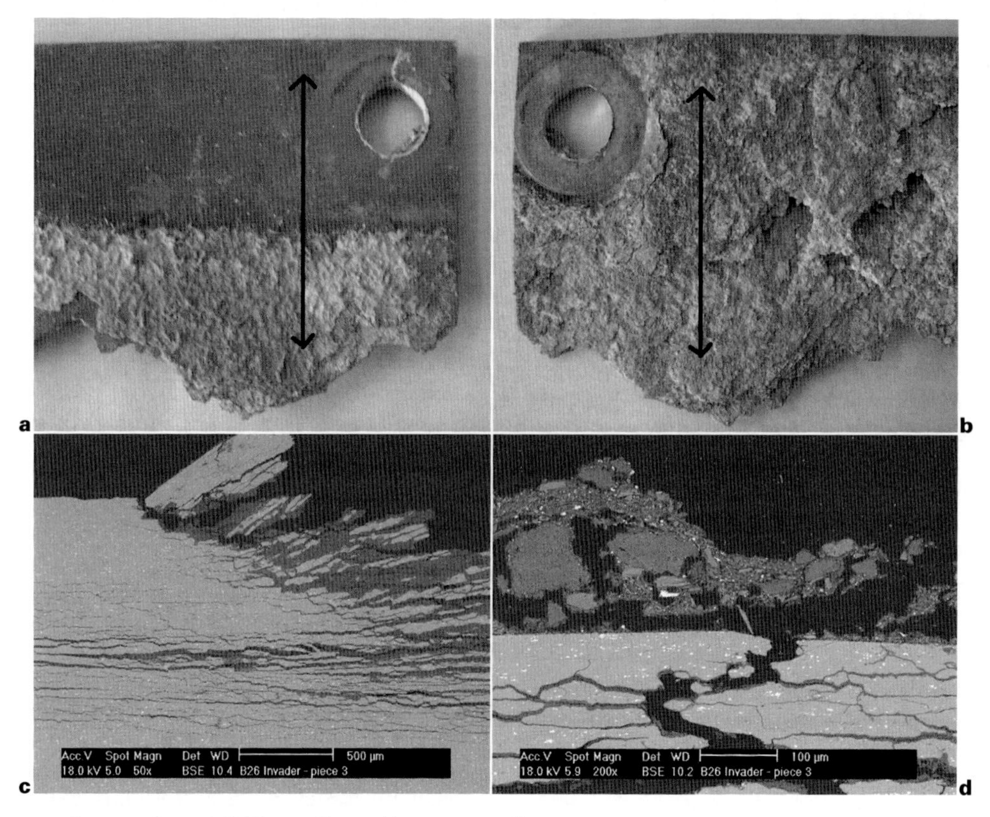

a, *b* macroscopic aspect; *c*, *d* SEM metallographic cross-sections
2 Example of lamellar corrosion: Vautour (spar)

a, *b* macroscopic aspect; *c*, *d* SEM metallographic cross-sections
3 Example of intergranular corrosion: Breguet 19 (ribbon of wing)

mineral corrosion inhibitors based on Cr compounds and mineral charge based alumininosilicate compounds (clays).

Concerning the corrosion products, elemental analysis by EDX and EPMA gives an approximate composition close to the Al/O=1 : 3 molar ratio on all samples. In general, they are amorphous or poorly crystallised as revealed by X-ray diffraction analysis. Some measurements by X-ray atomic spectroscopy, realised in Swiss Light Source synchrotron (Lucia line), have shown that the Al atoms in these amorphous compounds have a chemical environment close to the one in gibbsite $Al(OH)_3$.

The last kind of analysed sample comes from the Canadair aircraft (1966) and is displayed in Fig. 5. This sample is exempt of corrosion, but presents successive paint layers. Six paint layers are observed on this sample due to successive maintenance operations (Fig. 5*b*).

Reactivity study of 'Al alloys/corrosion layer' systems by electrochemistry

During several months, the humidity and temperature measurements performed inside the aircrafts exposed in large hall of the Air and Space Museum reveal important variation from 5 to 30°C for temperature and from 30 to 85% for the relative humidity. These atmospheric conditions provoke successive wet/dry cycles on metallic pieces. In some cases, it is possible to observe a complete wetting of some pieces during the condensation period.

Consequently, the second part of the authors work is devoted to the study of the reactivity of complex system, 'Al alloys/corrosion layer', by electrochemical measurements. The electrolytic medium is a corrosion reference

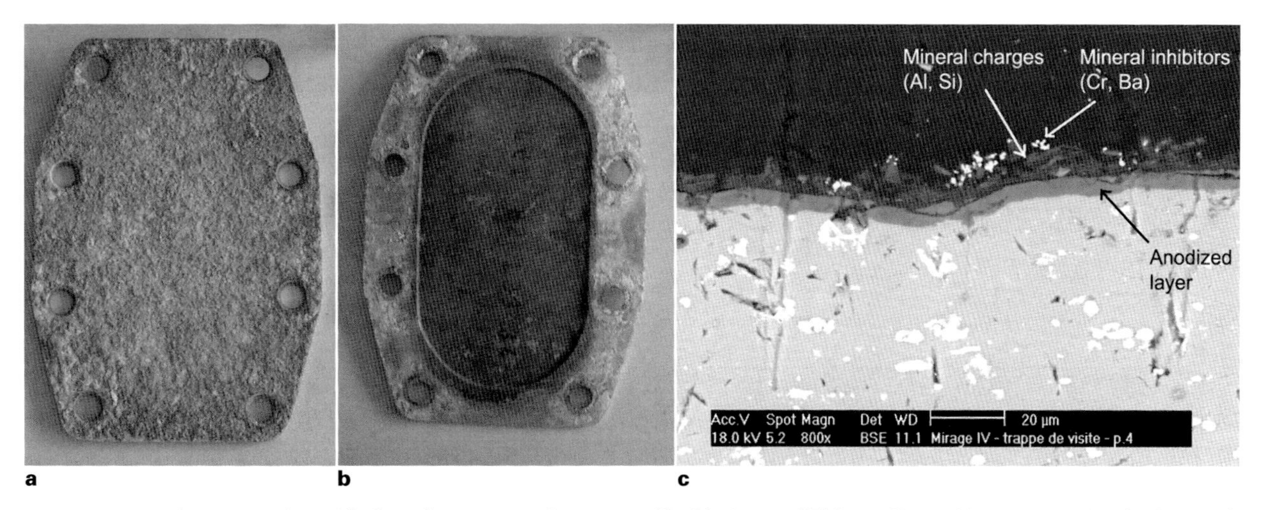

a macroscopic aspect of outside face; *b* macroscopic aspect of inside face; *c* SEM metallographic cross-section (inside face)
4 Hatch located under wing of Mirage IV aircraft

a macroscopic aspect; *b* SEM metallographic cross-section

5 Example of samples with paints: Canadair

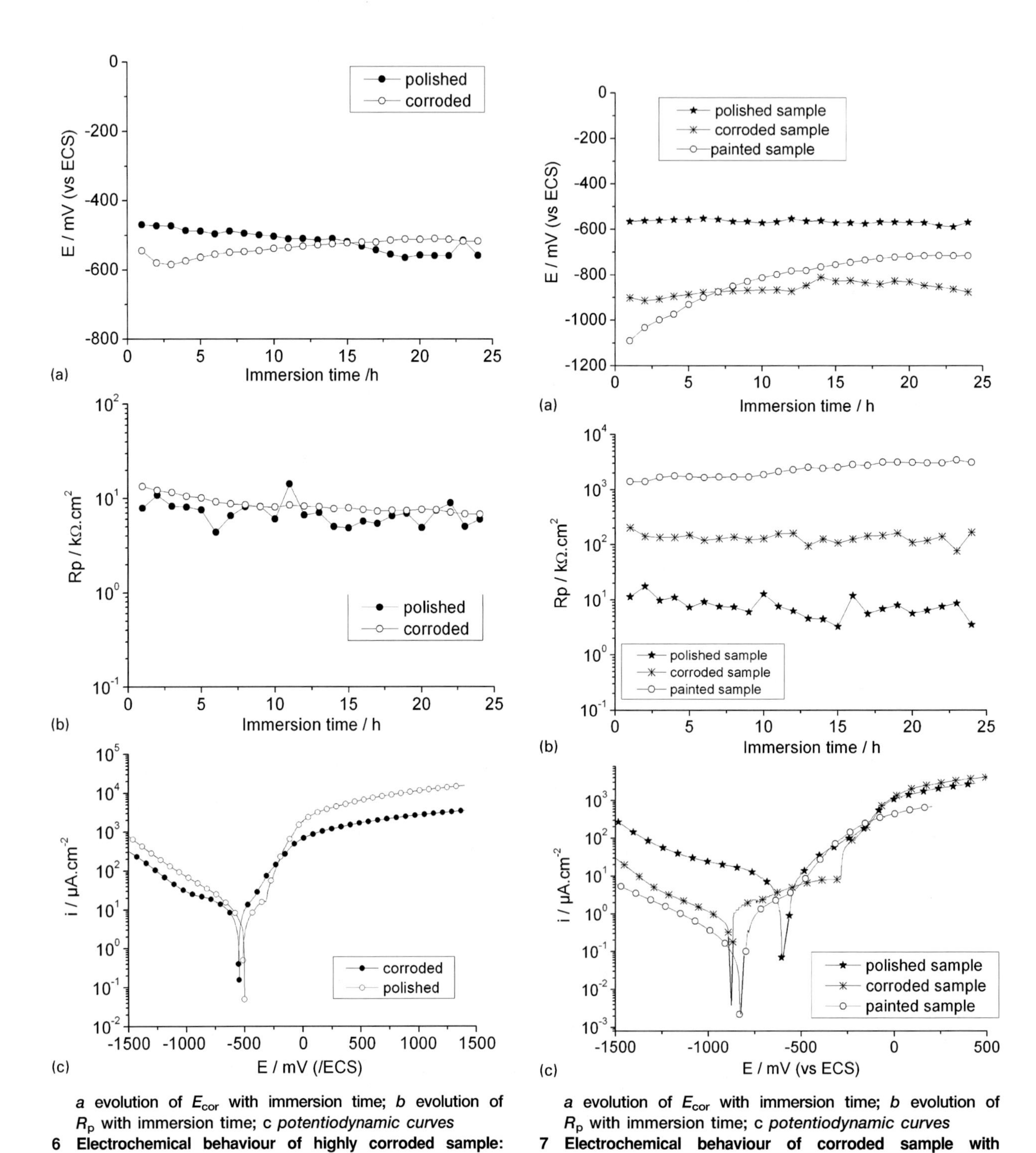

a evolution of E_{cor} with immersion time; *b* evolution of R_p with immersion time; c *potentiodynamic curves*

6 Electrochemical behaviour of highly corroded sample: outside face of Mirage IV hatch

a evolution of E_{cor} with immersion time; *b* evolution of R_p with immersion time; c *potentiodynamic curves*

7 Electrochemical behaviour of corroded sample with paint: fuselage sample of Short Sandrigam Bermuda

solution composed by 148 mg L^{-1} Na_2SO_4, 138 mg L^{-1} $NaHCO_3$ and 165 mg L^{-1} NaCl.[7] This electrolyte allows simulating the water layer on the surface of artefacts during the wet periods in atmospheric conditions.[8] For each sample, the measurements were carried out on polished samples (bare metal) and corroded samples (alloy/corrosion layer). The following electrochemical measurements were performed as follows:[9]

(i) measurements of corrosion potential E and polarisation resistance R_p (at 0·166 mV s^{-1}) during 24 h of immersion
(ii) recording of i=f(E) potentiodynamic curve at 1 mV s^{-1}.

On all samples, the results show two types of electrochemical behaviour in function of the sample: the highly corroded samples (with lamellar or internal corrosion) and the samples with paints and surface treatment residues.

The electrochemical behaviour of a highly corroded sample is illustrated by the measurements carried out on the hatch of the Mirage IV (Fig. 6). With or without the corrosion layer, the Al alloy has similar corrosion behaviour. The corrosion potential and polarisation resistance values and the potentiodynamic curves are quite similar (Fig. 6). This fact demonstrates that the corrosion layer on Al alloys have no effect on the corrosion rate of the metal.

In the case of samples taken on the fuselage of the Short Sandrigam Bermuda aircraft, three different surface states could be tested by electrochemistry on the same piece:

(i) samples with important corrosion product layer (corroded sample)
(ii) samples covered with paint (painted sample)
(iii) bare metallic sample (polished sample).

Figure 7 reveals that the bare sample is more active than that of the corroded and painted one. This behaviour is characterised by low R_p values on Fig. 7*b* and a high corrosion current density on Fig. 7*c*. Nevertheless, the corroded and painted samples is characterised by more cathodic values of corrosion potential and high values of polarisation resistance. Therefore, on these two samples, the corrosion layer containing paint residues on the surface has a corrosion inhibiting action, which is mainly due to the inhibition of the cathodic reaction, as can be clearly seen on potentiodynamic curves displayed on Fig. 7*c*. In fact, many mineral inhibiting compounds, present in paints and based on Zn and Cr, have a cathodic inhibiting action. Consequently, the authors can observe that, despite the fact that the organic polymer of paint is completely destroyed in places, these mineral inhibiting compounds are mixed to the corrosion products and have yet an important protective action.

The authors can also note that the bare metal of Mirage IV and Short Sandrigam Bermuda has the same corrosion behaviour, which is corroborated by the similar chemical analysis of the metal. In the two cases, the composition is very close to AA 2024 standard composition.[1]

Conclusions

Through the results obtained in this work, an overview of problems of long term corrosion of Al materials of heritage can be drawn up in the case of the aeronautic collections. Several conclusions can be made.

1. Corrosion morphology: the main corroded Al alloys in these conclusions are Al–Cu alloys on which very thick corrosion layers can be developed. This corrosion phenomenon is also characterised by an important intergranular corrosion inside the metallic pieces, which can lead to the crumbling of the metal and the complete disintegration of the metallic components. This phenomenon can be either very easily detectable or completely hidden. Therefore, the diagnosis and the expertise of the objects is often complex and can need several samplings and analysis. Nevertheless, the composition of the corrosion products is, in all cases, Al oxyhydroxides or hydroxides which are more or less crystallised.

2. Reactivity: whatever the manufacturing period of aircrafts, the Al–Cu alloys seem to have similar corrosion behaviour. The presence of thick corrosion layers has a poor influence on the corrosion rate of the bare alloy. Contrary to the case of long term corrosion of ferrous alloy, the Al corrosion layer does not contain oxidant species as Fe(III) compounds or conductive compounds as Fe_3O_4.[10]

The important result is that, in many cases, the corrosion layer can contain some mineral inhibiting compounds coming from paint residues, which can still have an important action of corrosion inhibition. In fact, the use of Zn or Cr based mineral compounds has been early used in paints as corrosion inhibitor.

In the practical point of view, the 'corrosion' diagnosis of the aeronautic collection is often uncertain and difficult because of the importance of the internal corrosion. The characterisation of the reactivity of these 'Al alloys/corrosion layer' complex systems in immersed conditions by electrochemical measurements seems to be not sufficient to completely conclude. Further works will be devoted to the study of hydration or swelling of Al corrosion products with humidity, which seems to be the cause of the crumbling and destroying of many pieces.

Acknowledgement

The authors thank the French Ministry of Culture for its funding in the framework of the National Research Program in conservation.

References

1. J. R. Davis (ed.): 'ASM specialty handbook: aluminium and aluminium alloys'; 1983, Materials Park, OH, ASM International.
2. P. Campestrini: *Corros. Sci.*, 2000, **42**, 1854.
3. E. Rocca and J. Steinmetz: *J. Electroanal. Chem.*, 2003, **543**, 153.
4. C. Bagnall: *Microsc. Microanal.*, 2009, **15**, 160.
5. F. Eckermann, T. Suter, P. J. Uggowitzer, A. Afseth and P. Schmutz: *Corros. Sci.*, 2009, **50**, (7), 2085.
6. G. Thompson and G. C. Wood: 'Anodic films on aluminum', in 'Treatise on materials science and technology', (ed. J. C. Scully), 1983, New York, Academic Press.
7. 'Standard test method for corrosion test engine coolants in glassware', D1384, ASTM, West Conshohocken, PA, USA, 1988.
8. E. Rocca, C. Caillet, A. Mesbah, M. François and J. Steinmetz: *Chem. Mater.*, 2006, **18**, (26), 6186–6193.
9. C. Georges, E. Rocca and P. Steinmetz: *Electrochim. Acta*, 2008, **53**, 4839–4845.
10. E. Rocca, C. Rapin and F. Mirambet: *Corros. Sci.*, 2003, **46**, 653–665.
11. J. Monnier, L. Legrand, L. Bellot-Gurlet, E. Foy, S. Reguer, E. Rocca, P. Dillmann, D. Neff, F. Mirambet, S. Perrin and I. Guillot: *J. Nucl. Mater.*, 2008, **379**, (1–3), 105–111.

In situ measurement of oxygen consumption to estimate corrosion rates

H. Matthiesen*[1] and K. Wonsyld[2]

This paper presents a novel non-destructive method to measure the actual corrosion rate of precorroded metal objects, such as historical and archaeological artefacts. The corrosion rate is estimated from the oxygen consumption of the objects, which is measured in a small volume of air encapsulated directly on the surface of the object. An optical method is used for the oxygen measurements, making it possible to measure through transparent materials such as glass. The method is tested on iron and copper samples in different environments using both new uncorroded metal and historical artefacts, which have thick corrosion scales from more than 50 years outdoors. The results show the following: the method has a good reproducibility; there is a good correspondence between oxygen consumption and weight loss; the corrosion rates of precorroded cast iron are significantly lower than the rates found for new steel samples, whereas corrosion rates for precorroded copper are equal to or higher than rates for new copper samples; and corrosion rates as low as 0·1 μm/year can be measured by the method.

Keywords: Atmospheric corrosion, Oxygen consumption, Corrosion rate, *In situ* measurement, Non-destructive

This paper is part of a special issue on corrosion of archaeological and heritage artefacts

Introduction

Corrosion rates within the field of atmospheric corrosion are normally estimated by measuring the changes in weight or resistance of well defined metal samples.[1,2] Extremely sensitive methods, such as the quartz microbalance or electrical resistivity probes, have been developed, where the changes can be followed continuously during the corrosion experiment. Other methods involve stripping of the corrosion products at the end of the experiment and quantification of metal ions in the stripping solution or weight loss of the metal sample.

However, when it comes to metal objects that already have a thick and possibly protecting layer of corrosion products, it becomes more difficult to estimate the corrosion rate: measuring the thickness of the corrosion layer or stripping of the corrosion products can only give an average corrosion rate for the whole lifetime of the object, and even then, this is only possible if no corrosion products have been lost to the environment. Measuring the weight increase over time is possible on smaller objects,[3] but for larger objects, it will normally require destructive sampling. Electrochemical impedance spectroscopy and chronoamperometry have been performed,[4] as well as characterisation of the reactivity of the corrosion products,[5] but all these studies required destructive sampling.

[1]Department of Conservation, National Museum of Denmark, Lyngby, Denmark
[2]Topsoe Fuel Cell, Lyngby, Denmark

*Corresponding author, email henning.matthiesen@natmus.dk

Within the field of cultural heritage, destructive sampling should be avoided as far as possible due to the value and rarity of the objects under study. Still the actual corrosion rates are very relevant in order to estimate the lifetime and need for protection of the objects. There have been some attempts to measure the corrosion rates non-destructively, focusing especially on *in situ* measurements of electrochemical impedance spectroscopy.[6] The results are promising and make it possible to compare different artefacts and treatments, even if it is still difficult to make estimates of the exact corrosion rates.

In this study, the corrosion rates are estimated based on the oxygen consumption, which is measured in a small volume of air encapsulated on the metal surface. Oxygen consumption has earlier been used for estimating corrosion rates,[7–9] but only on small test objects that fitted into a reaction chamber. The aim here is to develop a chamber that can be used directly on the surface of larger objects, thus avoiding destructive sampling (Fig. 1).

The method depends on a direct correlation between the oxygen consumption and the metal oxidation, which needs to be verified for the objects and environments that are studied. For wet–dry cycles, the two processes do not necessarily take place at the same time,[10] but still the overall oxygen consumption for the whole cycle should balance the metal oxidation. Corrosion under hydrogen development may decouple the oxygen consumption and metal corrosion, unless the H_2 developed is continuously reoxidised by O_2 within the chamber. Neither wet–dry cycles nor hydrogen development is investigated in detail in the present study, but their possible influence is checked by comparing the oxygen consumption with the weight loss of clean samples.

Published by Maney on behalf of the Institute
Received 11 December 2009; accepted 4 March 2010
DOI 10.1179/147842210X12710800383602

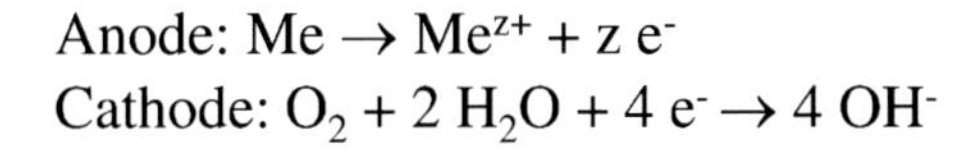

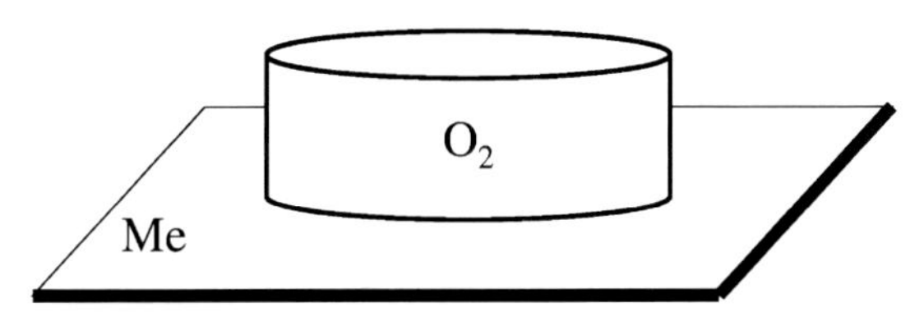

1 **Principle of method under study: oxygen concentration is measured in confined air volume above metal, created by gluing glass container on metal surface; as metal corrodes, oxygen concentration decreases**

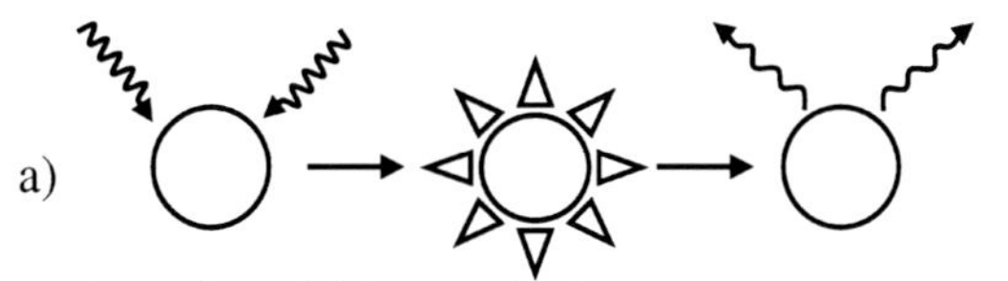

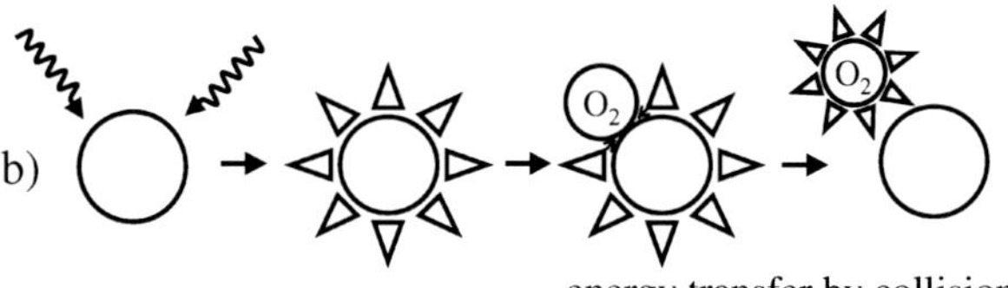

2 **Working principle behind optical oxygen electrodes: Fibox3LCD uses excitation wavelength of 505 nm and measures emission at 600 nm**

Methodology

All oxygen measurements are made with an optical method based on luminescence in which molecules are excited with light at one wavelength and emit the energy at another (Fig. 2*a*). Different oxygen sensitive molecules have been developed, where the presence of oxygen quenches the emission of light, and the oxygen concentration can be determined from the luminescence decay time or the intensity of the emitted light (Fig. 2*b*).[11,12] In this study, a Fibox3LCD oxygen meter from PreSens (www.presens.de), along with their sensor foil SF-PSt3-NAU-YOP has been used. The manufacturer makes a calibration of the sensor foil using different gas mixtures of N_2 and O_2. The calibration is checked frequently at our laboratory by measuring the response in nitrogen and in atmospheric air: this shows that the sensor foil and measuring equipment are very stable, showing insignificant drift even after a year. The response depends strongly on the temperature, for which the system compensates automatically via a built-in temperature sensor. The accuracy of the oxygen measurement is given as $\pm 1\%$ at 100% saturation, but higher inaccuracy may occur if there are temperature gradients between the oxygen sensor and the temperature sensor. More details on the system may be found in an earlier paper[13] where it is used for measuring oxygen consumption of smaller objects.

One of the advantages of an optical system is that it works through transparent materials such as glass (Fig. 3). This means that if a metal object (or just part of its surface) is encapsulated in glass along with some air and a piece of oxygen sensing foil, the oxygen concentration can be measured from the outside by sending and measuring light through the glass wall. There is no need to lead any electrodes or tubes through the chamber wall, which reduces the risk of leakage and allows the measurements to continue for prolonged periods.

Petri dishes or watch glasses are used as chambers, with a small sensor spot on the inside (Fig. 4*a* and *b*). The chamber is glued to the metal surface, and measurements are made by pointing an optical fibre from the Fibox3LCD unit towards the sensor spot. One of the main challenges is to find an adhesive that is sufficiently oxygen tight on all metal surfaces and that does not consume oxygen itself during or after the curing. An epoxy (Huntsman, base LY5138-2, catalyst HY5138, with a small amount of hydrolysed silica powder to adjust its viscosity) has been used in this study. Several different glues have been tested (including different polyurethanes, polyesters and epoxies), out of which epoxy had the best characteristics in terms of adhesion to the substrate and a low oxygen consumption and diffusion.[14] However, these tests were, by no means, all-embracing, and it is possible that there are other glues (including other epoxies) with even better properties.

Both Petri dishes (cylinder, Ø 4–5 cm; height 0·6–1 cm) and watch glasses (concave, Ø 4·7 cm; max. height, 0·4 cm) are used as chambers, which makes it possible to adjust the sensitivity of the method somewhat: the watch glass has a small air volume relative to the metal area, which makes it especially suitable for measuring low corrosion rates. For reference, some test specimens are encapsulated in glass vials closed with an airtight lid, instead of epoxy (Fig. 4*c*).

The corrosion experiments are made both on clean test panels and precorroded artefacts of Fe and Cu. Blind tests on glass plates instead of metal are carried out as well, to investigate if metal corrosion is the only process that consumes oxygen. The cleaned Fe and Cu samples are used for comparing the oxygen consumption with reference methods for corrosion rate measurements. The Fe samples are made from a low alloyed steel plate (St37), and the Cu samples are made from standard Cu plate, both of which are cathodically cleaned in a cyanide bath and rinsed in a fluoride

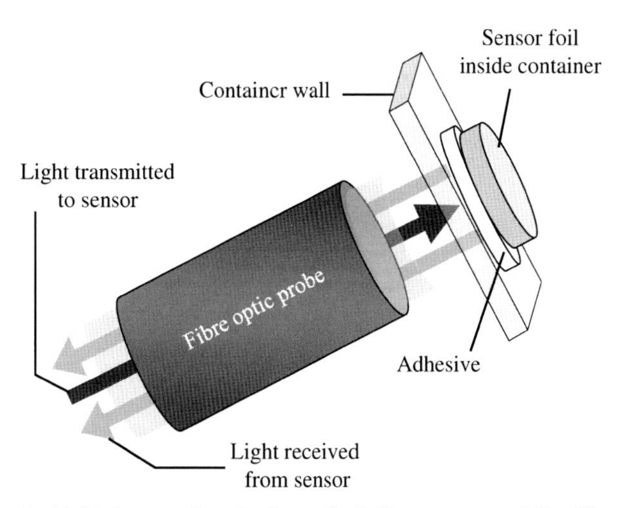

3 **Light is transferred via optical fibre: sensor foil with oxygen sensitive dye is placed inside container, and light is transferred and measured through transparent container wall**

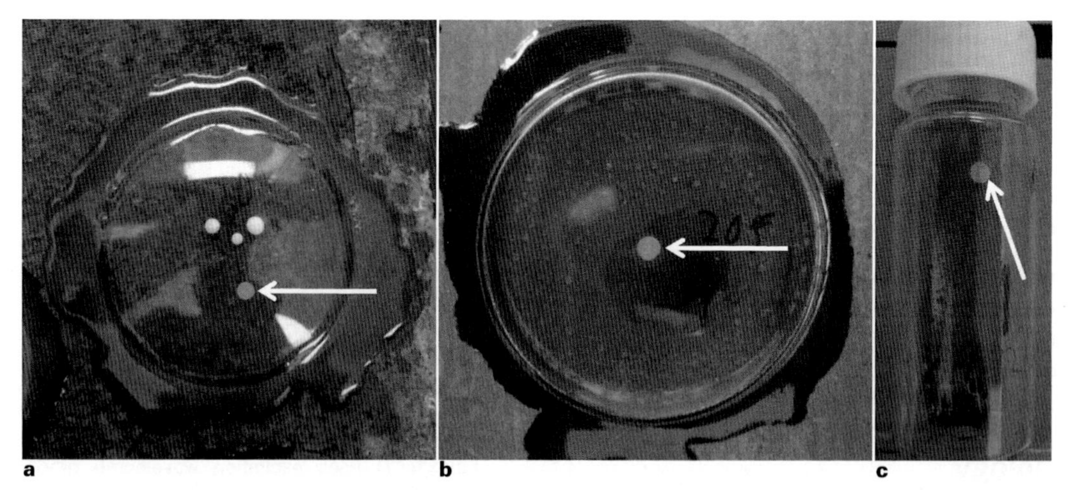

4 Encapsulation of samples. *a* watch glass glued to precorroded cast iron. The small white silica balls (ArtSorp) are used to buffer the relative humidity (RH) at 50%; these have in later experiments been replaced by Cl free silica gel. *b* Petri dish glued to the surface of precorroded Cu, water droplets inside the dish give an RH of 100%. *c* steel plate in a glass vial at 100% RH, the plate has been sprayed with a NaCl solution. The arrows mark round spots of oxygen sensor foil (Ø=3 mm) placed inside the containers

solution just before the experiment. The precorroded Fe sample is a >100 years' old window sill of cast iron, which is covered by an ~100 μm thick corrosion layer filled with cracks and inclusions. The precorroded Cu sample is a >50 years' old roof plate, which has been corroding in an urban atmosphere and which is covered by two distinct corrosion layers with a total thickness of 30–70 μm. These materials are used to test if the oxygen consumption measurement is sensitive enough to quantify corrosion rates on precorroded historical materials and also to get an idea of the spatial variation of the corrosion rate across the sample.

Three different environments are used for the tests: 100% relative humidity (RH), which is obtained by adding 0·17 mL ultra pure water inside the Petri dish, corresponding to an average water layer thickness on the metal surface of ~0·1 mm; 100% RH with NaCl, which is obtained by adding 0·17 mL of a NaCl solution with 1 g Cl/L; and finally, 50% RH, which is controlled by including a small amount of silica gel, which has been conditioned in a climate chamber beforehand (the product ArtSorp was used initially due to its high buffering capacity, but it contains corrosive LiCl and was later replaced by washed silica gel). Watch glasses (with a volume/area ratio of 0·2 cm) are used for experiments at 50% RH, whereas experiments at 100% RH are made with high Petri dishes (volume/area ratio of 1·0 cm) for Fe samples and low Petri dishes (volume/area ratio of 0·6 cm) for Cu samples. Reference tests with N_2 atmosphere are made to quantify the oxygen diffusion through possible leaks and through the bulk epoxy. All experiments are carried out in triplicate (apart from the blind tests) and at room temperature.

When the oxygen measurements are finished, the Petri dish or watch glass is removed from the sample. This is done by heating with a hairdryer, which makes the epoxy soft and easy to peel of. The corrosion products formed are characterised by Fourier transform infrared and X-ray diffraction on some of the clean samples, in order to check the oxidation state of the metal ions. Weight loss of the samples is determined by stripping of the corrosion products (quantitatively) using a stripping solution of diluted HCl (with 0·1% hexamethylene tetramine for Fe)[2] and weighing the samples. The metal content of the stripping solution is determined by flame atomic adsorption spectroscopy to give another measure for the amount of corrosion products.

Results

The oxygen concentration of the different samples has been measured for several months, and the changing concentrations are shown in Fig. 5. Only results from the first 100 days are shown, since after approximately four months, some of the actively corroding samples were no longer airtight, which was seen as a sudden increase in the oxygen concentration. Furthermore, some samples at 100% RH became dry after 0·5–1 year, possibly due to diffusion of water into the epoxy, which may take up ~10 wt-% water.[15,16] The samples in vials (Fig. 4*c*) showed less problems with blind consumption or leakage of oxygen, which after two years only amounted to a few per cent (data not shown).

For the clean metal samples, the oxygen measurements are stopped before the system becomes anoxic. The dishes and vials are opened, the corrosion products are cleaned off quantitatively and the weight loss is measured. The total oxygen consumption (mg O_2) is found as

$$\text{Oxygen consumption} = VC\Delta O_2/100\% \quad (1)$$

where V is the volume of air inside the dish/vial (cm^3), C is the initial concentration of oxygen (calculated by the ideal gas law to 0·276 mg cm^{-3} or 8·6 μmol cm^{-3} at 23°C) and ΔO_2 is the decrease in oxygen saturation (%) during the experiment.

The oxygen consumption is then compared to the weight loss of the samples (Fig. 6). For the Cu samples in vials, the oxygen concentration only decreased to 82–86% within two years, and the Cu samples from *in situ* measurements became untight before the weight loss measurements could be made.

Visual inspection and Fourier transform infrared analysis of the corrosion products formed on the clean Fe samples indicated a mixture of red lepidocrocite (FeOOH) and black magnetite (Fe_2O_3) when the experiment was finished and the weight loss was

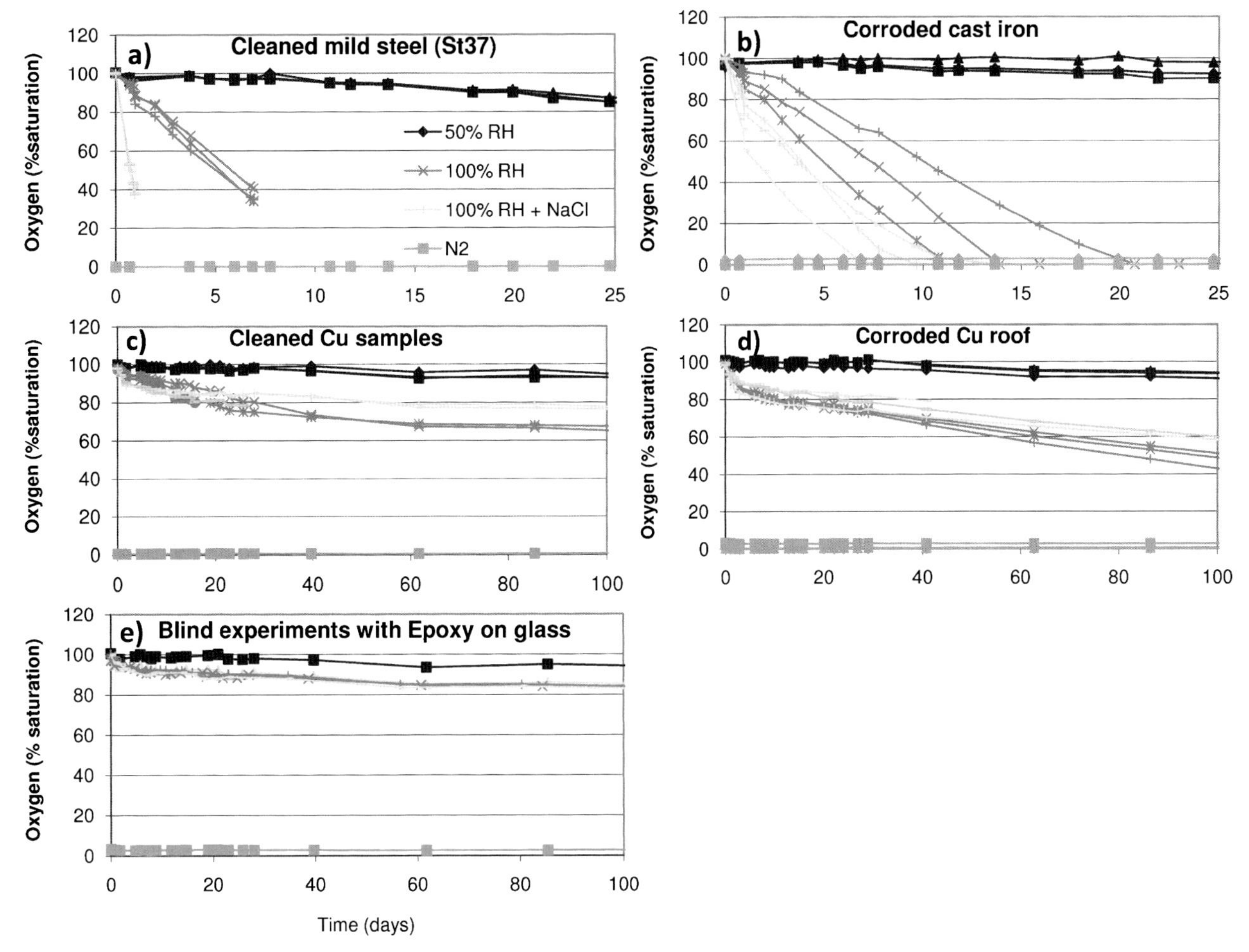

5 Oxygen concentration (per cent saturation) versus time (days) measured within watch glasses and Petri dishes fixed on surface of different test specimens: note different scale on *x* axis

determined. For the Cu samples, the weight loss measurements were carried out after two years, and here, no identification was made of the corrosion products; however, during the earlier stages (after four weeks), X-ray diffraction analysis showed black CuO for the samples corroding at 100% RH and reddish Cu_2O for the samples where NaCl was present.

Discussion

The purpose of this study was to validate if *in situ* measurements of oxygen consumption may be used to estimate the actual corrosion rate of precorroded artefacts. This requires discussion of several points.

Is system airtight?

The blind experiments with nitrogen filled containers (green curves in Fig. 5a-e) show no increase in oxygen concentration despite a concentration gradient between the container (0–3% oxygen saturation) and the ambient (100% oxygen saturation). This indicates that the system is airtight on all the tested surfaces and that oxygen diffusion through the epoxy is negligible. However, for the actively corroding samples, it was observed that, after approximately four months or more, the oxygen concentration inside the Petri dish or watch glass could suddenly increase probably because the corrosion had created a leak underneath the epoxy (not shown). The increase is very distinct, and it is easy to see when the system has failed and discard these results.

Is metal corrosion the only oxygen consuming process?

The blind experiments carried out on glass (Fig. 5*e*) show that, even for glass, there is a small consumption of oxygen, especially in the experiments at 100% RH: the oxygen saturation decreases to ~90% within 10 days (or a decrease of 1% per day), after which the rate is much slower (~0·05% per day). The blind experiments at 50% RH do not show the fast initial oxygen consumption, but a more constant decrease of ~0·05% per day.

This ‘blind consumption’ of oxygen needs to be compensated for, in order to find the amount of oxygen used for metal corrosion. It increases the uncertainty and detection limit of the method, and further work is necessary to make a system with a lower blind consumption. It is assumed that the oxygen consumed in these blind experiments is used to oxidise one or more components in the epoxy, and it seems that the consumption mainly takes place during curing of the epoxy at a high RH.

Are results reproducible?

The results from the *in situ* measurements of oxygen consumption on clean samples of Fe (Fig. 5*a*) and Cu (Fig. 5*c*) are highly reproducible, as there is very little difference between the three replicates for each environment. For comparison, the results for the precorroded cast Fe show a larger variation, which is probably due to the heterogeneous nature of the metal surface (Fig. 5*b*),

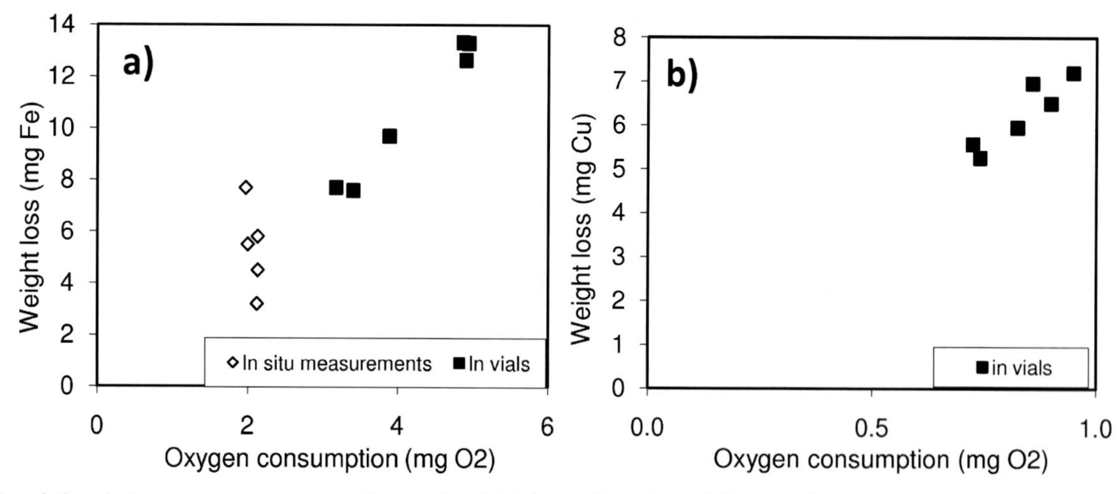

6 Correlation between oxygen consumption and weight loss of *a* cleaned Fe samples and *b* Cu samples: weight loss measurements were carried out after a few days for Fe samples, whereas it took two years before Cu samples were sufficiently corroded; oxygen consumption has been compensated for blind consumption measured in empty vial

giving a variable corrosion rate. It is thus concluded that the oxygen measurements themselves are reproducible, whereas the actual corrosion rates may vary across the precorroded samples.

Is there correlation between oxygen consumption and corrosion as measured by alternative methods?

Figure 6 shows that there is a good correlation between the oxygen consumption and the weight loss of the clean Fe and Cu samples in vials. For the *in situ* measurements on Fe, there is no clear correlation, but this is probably due to uncertainty of the weight losses: the total weight losses are only a few milligrams, and even if heating makes the epoxy soft and easy to peel of, it is difficult to remove it quantitatively down to the last milligram before the weighing. A better correlation is found when comparing the oxygen consumption and the amount of Fe in the stripping solution as measured by flame atomic adsorption spectroscopy (not shown).

The oxygen consumption measured for the precorroded samples cannot be compared to a weight loss, as a stripping solution would dissolve all corrosion products on the samples and not only the products formed during the oxygen measurements. Measurement of the weight increase is no good alternative due to the problems with removing the epoxy quantitatively and because both oxygen and adsorbed water may cause a weight increase in the samples. However, even without a formal proof, it is proposed that the oxygen consumption correlates to the metal corrosion for these precorroded samples, as it does for the clean metal samples.

Is oxygen reduction the only cathode reaction?

We are studying the reactions:

$$\text{anode } Me \rightarrow Me^{z+} + ze^-$$

$$\text{cathode } O_2 + 2H_2O + 4e^- \rightarrow 4OH^-$$

If these are the only reactions occurring, the weight ratio between the metal loss and the oxygen consumed would be

$$\frac{\text{Metal loss}}{\text{Oxygen consumption}} = \frac{4M_{Me}}{zM_{O_2}} \tag{2}$$

where M_{Me} and M_{O_2} is the molar mass of the metal and of oxygen respectively and z is the oxidation state of the metal in the corrosion products. It is thus necessary to know the composition of the corrosion products in order to calculate the theoretical weight ratio.

For Fe, the corrosion products on the cleaned samples consisted of a mixture of lepidocrocite (FeOOH) and magnetite ($Fe_2O_3.FeO$). For lepidocrocite, the oxidation state z is 3, giving a weight ratio between metal loss and oxygen consumption of 2·33. For magnetite, z is 8 : 3, giving a weight ratio of 2·62. For comparison, the Fe samples from glass vials (Fig. 6) gave metal loss/oxygen consumption ratios of 2·2–2·7.

For Cu, the corrosion products formed on the clean samples after four weeks contained CuO and Cu_2O, where the oxidation state z is 2 and 1 respectively (the corrosion products may have changed later, but still within the same range of oxidation states). This gives a weight ratio between metal loss and oxygen consumption of 4·0 and 7·9. For comparison, the weight ratios measured for the samples after two years vary between 7·1 and 8·1 (Fig. 6).

For both metals, there is a good correspondence between the measured and the theoretical weight ratio, which verifies that oxygen consumption can give a reasonable measure of the corrosion. This indicates that corrosion under hydrogen development is not quantitatively important under these conditions, or the H_2 developed is reoxidised by oxygen within the sample container. A more formal proof would require measurements of H_2 inside the containers or quantification of the exact amount of corrosion products in the different oxidation states.

Are measured corrosion rates realistic compared to other studies of atmospheric corrosion?

The oxygen consumption rate (OCR, mol cm^{-2} s^{-1}) can be found from the slope of the oxygen curves in Fig. 5

$$\text{OCR} = \frac{VC\Delta O_2/\Delta t}{A \times 100 \times 24 \times 3600} \tag{3}$$

where V is the volume of air (cm^3), C is the initial concentration of oxygen (mol cm^{-3}), $\Delta O_2/\Delta t$ is the slope of the oxygen curve (%sat/day), A is the area (cm^2) and

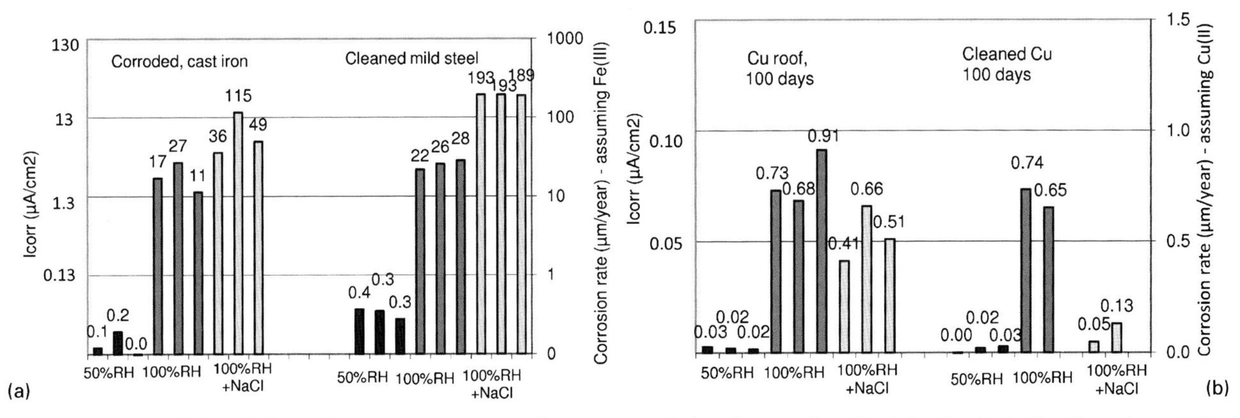

7 i_{corr} values calculated from the oxygen consumption measured *in situ*, as described in the text. For the cleaned Cu samples only two results are given for the samples at 100% RH, as the third sample was stopped already after four weeks to analyse the corrosion products. For recalculation to corrosion rates (μm/year), it is assumed that Fe is oxidised to Fe(III) and Cu to Cu(II). Numbers above the columns show the corrosion rate (secondary *y* axis). Note logarithmic scale for Fe samples

the constants 24×3600 s day^{-1} and 100 (%) are included to convert the number to mole per square centimetre per second. The slope of the oxygen curves from all *in situ* measurements have been estimated using linear regression and corrected for the blind consumption in order to compensate for the amount of oxygen used to oxidise the epoxy. For non-linear curves, only the first part of the curve (down to ~60% oxygen) is used in the regression, and no attempts have been made at this stage to make more detailed kinetic expressions from the shape of the curves.[7] The OCR has then been calculated for each experiment using the actual *V/A* ratio of the different Petri dishes and watch glasses.

The i_{corr} (A cm^{-2}) can be calculated as

$$i_{corr} = 4\mathrm{OCR} \cdot \mathrm{F} \quad (4)$$

where F is Faradays constant (96 485 C mol^{-1}) and 4 is the amount of electrons transferred to one oxygen molecule. The results are given in Fig. 7 (left *y* axis).

The i_{corr} values may be recalculated to a corrosion rate (d*s*/d*t*, μm/year) using the formula

$$\frac{ds}{dt} = \frac{i_{corr} M_{Me} \times 365 \times 24 \times 3600 \times 10000}{zF\rho} \quad (5)$$

where M_{Me} is the molar weight of the metal (g mol^{-1}), z is the oxidation state of the metal, F is Faraday's constant (96 485 C mol^{-1}), ρ is the density of the metal (g cm^{-3}) and the constants $365 \times 24 \times 3600$ s/year and 10 000 μm cm^{-1} are included to convert the number to micrometre per year. The only unknown is z, which may vary, but if it is assumed (for simplicity) that Fe is oxidised all the way to Fe(III) and Cu to Cu(II), it gives the corrosion rates shown in Fig. 7 (right, *y* axis): if the metals are only oxidised to Fe(II) and Cu(I), the rates would increase correspondingly.

The magnitude of these corrosion rates seems realistic: ISO 9223[17] describes the corrosivity of different atmospheres based on the time of wetness, the content of SO_2 and the deposition of Cl^-. The standard distinguishes between five corrosivity classes, of which the experiments at 50% RH is estimated to be class 1 (very low), at 100% RH class 3 or 4 (medium to high) and at 100% RH with NaCl class 5 (very high). The corresponding corrosion rates given by the standard for carbon steel are <1·3 μm/year for class 1, 25–50 μm/year for class 3, 50–80 μm/year for class 4 and 80–200 μm/year for class 5. This corresponds well with the results in Fig. 7. For Fe, it is noted that the precorroded samples corrode slower than the clean samples, which may either be due to thick protecting corrosion layers on the former or the difference in material (cast iron versus mild steel).

For the Cu samples, ISO 9223[17] gives corrosion rates of <0·1 μm/year for class 1, 0·6–1·3 μm/year for class 3, 1·3–2·8 μm/year for class 4 and 2·8–5·6 μm/year for class 5. The results from this study correspond to classes 1–3, i.e. slightly lower than expected. As for the precorroded samples, Kreislova *et al.*[18] give the long term corrosion rates for Cu roofs to 0·1–1 μm/year for moderately polluted urban atmospheres and 1–3 μm/year in heavily polluted atmospheres. Again, this corresponds well with the results in Fig. 7. Surprisingly, the NaCl does not seem to increase the corrosion rate, but on the contrary decreases it, and the precorroded samples seem to corrode at the same rate or faster than the clean samples. However, the oxygen consumption for some of the Cu samples is close to the blind consumption, which increases the uncertainty of the numbers.

The results in Fig. 7 demonstrate the versatility of the method, as corrosion rates spanning over three orders of magnitude (from <0·1 μm/year for the Cu samples at 50% RH up to 200 μm/year for Fe at 100% RH with salt) are measured in these experiments.

Conclusions

Overall, the method is considered promising for measuring the corrosion rate of precorroded artefacts, which is notoriously difficult by other means. The reproducibility is very good, and for clean samples, there is a good correlation to the results from standard methods (weight loss). Corrosion rates between <0·1 and 200 μm/year have been measured in this study, i.e. a range of 3 orders of magnitude. The sensitivity and accuracy of the method is limited by the oxygen consumption of the epoxy used, but if this can be solved, it should be possible to measure even lower corrosion rates. One of the main shortcomings of the current design is that it is not possible to simulate a full wet–dry cycle, as the measurements are carried out at fixed RH values. Even with this limitation, the method may give realistic corrosion rates for museum objects, as these are normally stored at fixed RH values.

Within the field of cultural heritage, it is important not to damage the objects, and the sensors can be easily removed by heating the epoxy with a hairdryer. It is possible to make repeated measurements on the same object and, for instance, compare the corrosion rate of an object before and after conservation treatment. The method is currently used at the National Museum of Denmark to compare the effect of different coatings for protection of industrial heritage.[19]

Acknowledgements

Professor Per Møller from the Danish Technical University and Lars Vendelbo Nielsen from MetriCorr are acknowledged for their comments on the method and its potential use. The Danish Directorate for Cultural Heritage is acknowledged for the financial support to buy the oxygen measuring equipment (grant no. 2003-3321/10101-0039).

References

1. C. Leygraf and T. E. Graedel: 'Atmospheric corrosion'; 2000, New York, Wiley.
2. 'Standard practice for preparing, cleaning and evaluating corrosion test specimens', G1-90, ASTM, Philadelphia, PA, USA, 1990.
3. L. Maréchal, S. Perrin, P. Dillmann and G. Santarini: in 'Corrosion of metallic heritage artefacts: investigation, conservation and prediction for long-term behaviour', (ed. P. Dillmann, *et al.*), 131–151; 2007, Cambridge, Woodhead Publishing Ltd, European Federation of Corrosion Publications.
4. E. Pons, C. Lemaitre, D. David and D. Crusset: in 'Corrosion of metallic heritage artefacts: investigation, conservation and prediction for long-term behaviour', (ed. P. Dillmann, *et al.*), 77–91; 2007, Cambridge, Woodhead Publishing Ltd, European Federation of Corrosion Publications.
5. J. Monnier, L. Legrand, L. Bellot-Gurlet, E. Foy, S. Reguer, E. Rocca, P. Dillmann, D. Neff, F. Mirambet, S. Perrin and I. Guillot: *J. Nucl. Mater.*, 2008, **379**, 105–111.
6. P. Letardi and G. Luciano: 'Use of electrochemical techniques in metal conservation', Proc. Conf. Metal '07, (ed. C. Degrigny, *et al.*), 44–50; 2007, Amsterdam, Rijksmuseum Amsterdam.
7. L. Caceres, L. Herrera and T. Vargas: *Corrosion*, 2007, **63**, 722–730.
8. L. Gråsjö, G. Hultquist, Q. Lu and M. Seo: *Mater. Sci. Forum*, 1995, **185–188**, 703–712.
9. M. Stratmann: *Corros. Sci.*, 1987, **27**, 869–872.
10. M. Stratmann and J. Müller: *Corros. Sci.*, 1994, **36**, 327–359.
11. R. N. Glud, J. K. Gundersen and N. B. Ramsing: in 'In situ monitoring of aquatic systems: chemical analysis and speciation', (ed. J. Buffle and G. Horvai), 19–73; 2001, Chichester, Wiley.
12. F. Guillaume, K. Greden and W. H. Smyrl: *J. Electrochem. Soc.*, 2008, **155**, (8), 213–219.
13. H. Matthiesen: *Stud. Conserv.*, 2007, **52**, 271–280.
14. J. A. Brydson: 'Plastics materials'; 1995, Oxford, Butterworth-Heinemann.
15. M. Al-Harthi, K. Loughlin and R. Kahraman: *Adsorption*, 2007, **13**, 115–120.
16. S. Popineau, C. Rondeau-Mouro, C. Sulpice-Gaillet and M. E. R. Shanahan: *Polymer*, 2005, **46**, 10733–10740.
17. 'Corrosion of metals and alloys – corrosivity of atmospheres – classification', ISO 9223, ISO, Geneva, Switzerland, 1992.
18. K. Kreislova, D. Knotkova and V. Cihal: in 'Corrosion of metallic heritage artefacts: investigation, conservation and prediction for long-term behaviour', (ed. P. Dillmann, *et al.*), 263–271; 2007, Cambridge, Woodhead Publishing Ltd, European Federation of Corrosion Publications.
19. Y. Shashoua and H. Matthiesen: *Corros. Eng. Sci. Techn.*, 2010, **45**, 357–361.

Protection of iron and steel in large outdoor industrial heritage objects

Y. Shashoua* **and H. Matthiesen**

Preservation of large outdoor cultural objects containing iron and steel for future generations involves protecting them with a coating. To meet the ethics of the conservation profession and the typical budget, coatings should maintain the original appearance and significance of objects as well as require no maintenance for at least three years. Eighteen coatings, waxes and oils, which met the project's requirements based on peer reviewed research publications, were compared on an equal basis in practice for their potential to inhibit corrosion of large industrial objects. They were applied to test Q panels at equal dry film thicknesses and evaluated by mechanical, atmospheric and accelerated corrosion tests. Uncoated Q panels corroded significantly more rapidly than all coatings, suggesting that any treatment is better than nothing. Interim results suggest that the coatings which clearly inhibit corrosion, as shown by salt spray tests after 504 h, atmospheric corrosion for six months and oxygen consumption were Cosmoloid H80, Dinitrol Car/4941, LPS3, Rustilo 3000, SP400 and VpCI-386.

Keywords: Corrosion, Coating, Wax, Mechanical testing, Oxygen consumption, Atmospheric corrosion

This paper is part of a special issue on corrosion of archaeological and heritage artefacts

Introduction

In 2007, the National Museum of Denmark initiated a three-year, multidisciplinary research project into the conservation of industrial objects for future generations. The project focuses on the protection of industrial objects containing iron or steel. Such objects are large and include machinery, transport vehicles and military equipment. Protecting cultural objects usually involves controlling the temperature, relative humidity and light levels in which they are stored or displayed to minimise the rate of any degradation or corrosion reactions. However, because of their size and immobility, industrial objects are either found outdoors or indoors but without control of relative humidity, temperature or light levels. The purpose of the project was to identify a surface treatment, which would both protect iron and steel industrial objects from corrosion and comply with the ethics of conservation.

After discussion with professionals concerned with the study and care of industrial heritage objects in various organisations, including English Heritage, German Mining Museum and Danish Technical Museum, a list of ideal requirements for coatings was compiled:

(i) treatments should last for a minimum of three years outdoors or indoors but without control of relative humidity, temperature or light levels without requiring further maintenance

(ii) treatments should allow retention of the original appearance, function or cultural significance of objects

(iii) treatments should be removable or allow future retreatment

(iv) products should be commercially available

(v) treatments should be harmless to both operator and the environment.

Research strategy

An extensive investigation of peer reviewed literature published since 2000 and commercial sources of protection systems for outdoor steel, including coatings, oils, waxes and corrosion inhibitors, was made. Approximately 100 publications were deemed relevant to the present project. Only those materials that performed best in each study were selected, resulting in a total of 40. Their properties were compared with the list of ideal requirements developed for the project. Those which did not meet the health and safety requirements according to literature or were not commercially available were immediately discarded.

Information about longevity was not available for all protection systems. Technical data provided by the manufacturer of VpCI-386 acrylic primer/topcoat suggested that it protected surfaces for 5–10 years[1] while Ship-2-Shore Industrial, a liquid barrier coating, was claimed to protect surfaces for 5–20 years.[2] By contrast, Rustilo DWX 22 was claimed to last only one to two months outdoors, so it was not investigated further.[3] Eighteen materials seemed most likely from literature to meet all the ideal requirements and were evaluated

Department of Conservation, National Museum of Denmark, I.C. Modewegsvej, Brede, Kongens Lyngby, Denmark

*Corresponding author, email yvonne.shashoua@natmus.dk

Published by Maney on behalf of the Institute
Received 14 December 2009; accepted 29 January 2010
DOI 10.1179/147842210X12710800383648

Table 1 Surface treatments evaluated in investigation

Surface treatments	Major components of surface treatments	Appearance when applied	Suppliers
Corroheat 4010	Wax + corrosion inhibitor + solvent	Colourless + transparent + matte	EFTEC Aftermarket GmbH, Pyrmonter Str. 76, D-32676 Lugde, Germany
Cosmoloid H80	Microcrystalline waxes + hydrocarbon solvents	Colourless + transparent + matte	Kremer Pigments GmbH & Co KG (details as for Renaissance wax)
Dinitrol Car/4941	Wax + corrosion inhibitor + hydrocarbon solvent	Black + opaque + matte	EFTEC Aftermarket GmbH (details as for Corroheat 4010)
Frigilene	Cellulose nitrate + acetone	Colourless + transparent + medium gloss	Conservation By Design Ltd (details as for Paraloid B72)
Incralac	Acrylic + corrosion inhibitor + antioxidant + solvent	Colourless + transparent + medium gloss	Conservation By Design Ltd (details as for Paraloid B72)
LPS3	Corrosion inhibitor + wax + solvent	Brown + transparent + matte	ITW Chemical Products Scandinavia, Priorsvej 36 DK-8600 Silkeborg, Denmark
Paraloid B72	Acrylic	Colourless + transparent + high gloss	Conservation By Design Ltd, Timecare Works 5 Singer Way Woburn Rd, Ind. Estate Kempston, Bedford MK42 7AW UK
Perfluorodecyl iodide in Paraloid B72	Wetting additive-perfluoro	Colourless + transparent + medium gloss	Alfa Aesar GmbH & Co KG, Zeppelinstrasse 7, 76185 Karlsruhe, Germany
Poligen ES91009	Wax + surfactant + water	Colourless + transparent + high gloss	BASF Aktiengesellschaft Carl-Bosch-Str.38, 67056 Ludwigschafen, Germany
Renaissance wax	100 g Cosmoloid H80: 25 g polyethylene wax + 300 mL hydrocarbon solvent	Colourless + transparent + crazed	Kremer Pigments GmbH & Co KG, Haupstrasse 41–47, D-88317 Aichstetten, Germany
Rustilo 2000	Wax + corrosion inhibitor	Brown + transparent + matte	Kemi Service A/S Bugtattivej 15, DK-7100 Vejle Denmark
Rustilo 3000	Bitumen + wax + corrosion inhibitor	Brown + transparent + matte	Kemi Service A/S (details as for Rustilo 2000)
Ship-2-Shore Industrial	Unknown + water	Colourless + transparent + high gloss	Ship-2-Shore, Box 48205, Victoria, BC V8Z 7H6, Canada
SP400	Wax + corrosion inhibitor	Brown + transparent + matte	EFTEC Aftermarket GmbH (details as for Corroheat 4010)
Tectyl 506 rust preventative	Wax + corrosion inhibitor	Brown + transparent + matte	Eurodeal Autoparts A/S, Stamholmen 111, DK-2650 Hvidovre, Denmark
Tectyl Glashelder/Klar spray	Acrylic + corrosion inhibitor + solvent	Colourless + transparent + high gloss	Eurodeal Autoparts A/S (details as for Tectyl 506 rust preventative)
Tromm III	Waxes [TeCerowax 30201 and TeCerowax 30401 (1 : 1)]	Colourless + opaque	Th.C. Tromm GmbH, PO box 620168, D-50694 Cologne, Germany
VpCI-386 acrylic primer/topcoat	Acrylic + corrosion inhibitor + water	Colourless + transparent + high gloss	HITEK Electronics Materials Ltd. 15, Wentworth Road, South Park Industrial Estate, Scunthorpe, North Lincolnshire DN17 2AX, UK
None (bare Q-panel)	Low carbon, cold rolled steel panels with a dull finish and dimensions 10 × 15 × 0·8 mm	Matte	Bjoern Thorsen Oesterfaelled Torv, 14 Copenhagen OE DK-2100 Denmark

(Table 1).[4–6] Materials either contained corrosion inhibitors, protected metals mainly by forming a barrier against water or protected by forming a barrier against oxygen and atmospheric pollutants.

It is clear from the number of references that protection of iron and steel surfaces is a highly active area of research both in the conservation and commercial fields. However, selection of a suitable treatment to protect industrial objects from literature sources alone is a minefield. The three key factors which prevent inter-research comparison are variation in dry film thickness, use of non-standard evaluation techniques and selection of appropriate reference coatings.

Because many publications stated that protection systems had been applied according to manufacturer's instructions, a trial was made to determine the variation in film thicknesses achieved using such an approach. Q panels type R are low carbon, cold rolled steel panels with a dull finish and dimensions 10 × 15 × 0·8 mm and are the standard test substrates used to evaluate commercial coatings for vehicles. Q panels were used as supplied immediately after rinsing with acetone to remove the anticorrosion agent they were supplied with. The dry film thickness of each treatment after application according to manufacturers' guidelines was determined using a Fischer Dualscope MP4C probe. Mean thicknesses varied from <1 μm (Renaissance wax) to 50 μm (Dinitrol Car/4941).

The selected treatments each had met their respective research projects' criteria under various application and testing regimes. To compare their performances equally, it was decided to apply surface treatments so that they all attained 20–25 μm dry film thickness, despite the fact that this thickness may not be that recommended for all the coatings investigated. Twenty-five micrometres was selected as a halfway point between the thinnest and thickest films and because films thinner than 25 μm are known to offer poor protection to car bodies exposed to water, oxygen, salt and abrasion.[7] Brushing was selected unless the coating was supplied in an aerosol can, in which case it was sprayed. Brushing is a slow process, but the applied film thickness can be readily controlled and the health protection equipment required for brushing is generally less than that required for spraying. Each surface treatment protection system was applied using a 1 in. (2·5 cm) flat bristle brush to three, freshly cleaned Q panels. They were allowed to dry horizontally for seven days at 18–20°C and 35–40% relative humidity before evaluation.

Evaluation of protection systems

The standard techniques used to evaluate industrial coatings were applied to the newly applied coatings in this project. They include appearance, adhesion, hardness and resistance of surfaces to corrosion.

Appearance

Surface treatments were either colourless and transparent, brown and transparent or black and opaque (Table 1). All the brown coatings contained corrosion inhibitors. Treatments based on waxes did not form cohesive films, and brushstrokes and crazing were visible with the naked eye. The wetting agent, perfluorodecyl iodide (PI), imparted a milkiness to Paraloid B72 on drying, which indicated that they were poorly compatible. VpCI-386 acrylic primer/topcoat, Tectyl Glashelder and Poligen ES91009 produced highly glossy, colourless, transparent surfaces which would suit, from a cosmetic perspective, the surface appearance of polished new steel. Rustilo 3000 and Dinitrol Car/4941 formed black coatings, which were likely to protect surfaces from ultraviolet radiation more effectively than transparent films and would suit objects that were originally black. Ship-2-Shore Industrial produced a non-drying, tacky surface, which was readily marked and dust adhered readily to it.

Hardness

The hardness of surface treatments gives an indication of how readily they can be marked or abraded by airborne particles or during handling. Thumbnail hardness is a very practical and rapid procedure for comparing hardness of painted components.[8] Poligen ES91009, Frigilene and Incralac produced the hardest films, while Ship-2-Shore Industrial, LPS3 and Rustilo 3000 were readily marked and could not be handled without sustaining damage (Table 2). When applying these results to real use, it should be considered that some of the coatings would be applied as thicker films

Table 2 Performance of surface treatments

Surface treatment	Hardness of coating, marked with thumbnail?	Adhesion to Q panel (per cent of coating remaining on panel)	Percentage surface covered by corrosion after 504 h salt spray test, %	Visible changes in Danish weather for six months	Rate of corrosion by oxygen consumption after 70 days, μm/year
None (bare panel)			100	Corroded	130
Corroheat 4010	Yes	97	0	Discoloured	2·7
Cosmoloid H80	No	85	0	No change	<0·1
Dinitrol Car/4941	Yes	100	0	No change	0·7
Frigilene	No	100	50	Corroded	0·1
Incralac	No	90	8	Corroded	<0·1
LPS3	Yes	40	0	No change	0·1
Paraloid B72	No	86	8	Corroded	0·2
Paraloid B72 with 1% perfluorodecyl iodide	No	95	50	Corroded	0·4
Poligen ES91009	No	100	50	Discoloured/corroded	16·6
Renaissance wax	Yes	98	50	Corroded	1·9
Rustilo 2000	Yes	100	8	Corroded	...
Rustilo 3000	Yes	100	1	No change	0·4
Ship-2-Shore Industrial	Yes	Liquid, unable to measure adhesion	50	Corroded	1·9
SP400	Yes	95	0	No change	0·2
Tectyl 506 rust preventative	Yes	100	0·5	Discoloured	1·5
Tectyl Glashelder/Klar spray	Yes	100	50	Corroded	1·5
Tromm III	No	99	50	Corroded	3·5
VpCI-386 acrylic primer/topcoat	No	100	1	No change	0·1

1 Three panels each of coating were subjected to atmospheric corrosion testing. The panels were mounted on two frames in June 2009, one of which is shown to the left. The same panels show discolouration and the formation of corrosion by December 2009 (right)

and others as thinner ones than 20–25 μm when following manufacturers' guidelines.

Adhesion of protection systems to Q panels

A rigorous and simple test used to compare and quantify adhesion of protection systems to metal substrates is used by the coatings industry and was applied here. The crosshatch adhesion test is described in ASTM D3359-08 'Standard test methods for measuring adhesion by tape test' and involved cutting a grid of 100 squares into the coated panels. A pressure sensitive tape (3M) was applied over the grid and immediately peeled away. The number of whole squares remaining on the panel gave the percentage value of adhesion.

Most of the surface treatments adhered firmly to bare Q panels (Table 2). However, less than half of the original soft waxy film produced by LPS3 (40%) remained after applying the tape. Cosmoloid H80 wax and Paraloid B72 did not adhere as well as many of the other treatments (85 and 86% respectively), although adhesion of Paraloid B72 was improved by the addition of PI.

Salt spray test

Three panels for each coating underwent standard salt spray exposure as described by DS/EN ISO 9227 at the Danish Technological Institute. Non-coated back sides of panels were protected with an alkyd paint. Panels were tilted 20° vertical and exposed to 5% sodium chloride solution at 35°C for 504 h. During and after exposure, panels were visually examined for the percentage of rust present on the midsection as described by DS/EN ISO 4628-3, 2004.

Only Corroheat 4010, Cosmoloid H80, Dintrol Car/4941 and LPS3 showed no rust after 504 h (Table 2). Paraloid B72 with PI developed large blisters during salt spray, and the coating had lifted away from the panel. Rust covered between 1 and 50% of the areas of panels of all other coatings.

Atmospheric corrosion testing

Three panels for each coating and three uncoated panels were exposed to Danish weather (average temperature of 16·8°C, total rainfall of 376 mm and total hours sunshine of 1028) in real time for six months between June and December 2009. Exposure will continue for three years. Two exposure stands were constructed from galvanised steel according to the design described in ISO 8565 'Metals and alloys – atmospheric corrosion testing – general requirements for field tests specifications'. Nylon screws and spacers prevented test panels from coming into contact with the metal frame. Exposure stands were placed in a south facing position, 1·5 m above the ground in an open courtyard. Photographs of the stands were taken every Tuesday at 2 p.m. to document changes in appearance (Fig. 1).

Uncoated panels showed corrosion after exposure for only two days. Tectyl Glasheder, Paraloid B72 with PI, Frigilene and Rustilo 2000 discoloured after 30 days, most likely due to photodegradation in the presence of high ultraviolet levels. After 90 days, Corroheat 4010 and Ship-2-Shore Industrial also changed colour, and rust spots were visible. Tromm III, Paraloid B72 and Poligen ES91009 yellowed after 180 days, and patches of rust were visible.

Oxygen consumption

The corrosion rate of the panels was estimated from their oxygen consumption.[9] This method was chosen because it may also be applied to precorroded Q panels and artefacts in the next step of the project. The oxygen consumption inside Petri dishes glued to the surfaces of the coated Q panels was determined. Each Petri dish contained a sodium chloride solution (1 g Cl^- L^{-1}), which created a corrosive environment on the surfaces of panels. All oxygen measurements were made with an optical method using a Fibox3LCD oxygen meter from PreSens (http://www.presens.de) and their sensor foil SF-PSt3-NAU-YOP. Background measurements of the oxygen consumption by the various coatings applied to glass plates instead of metal were carried out to investigate whether corrosion of the Q panels was the only process that consumed oxygen. The blind consumption was subtracted from the results for the Q panels, and a corrosion rate in micrometre/year was calculated by assuming that the iron was oxidised to Fe(III) (Table 2). Measurements were carried out for 70 days.

No corrosion at all is unrealistic to achieve, so it was decided that an upper corrosion rate of 1 μm/year would be considered the same as none. Uncoated Q panels corroded at a rate of 130 μm/year, which was the highest rate measured. Panels treated with Poligen ES91009 exhibited the highest corrosion rate by treated metal at 16·6 μm/year, while Corroheat 4010, both Tectyl products, Ship-2-Shore Industrial, Renaissance wax, Tromm III, Tectyl 506 rust preventative and Tectyl Glashelder showed rates between 1 and 4 μm/year. All the other coatings produced corrosion rates of <1 μm/ year.

Discussion

The experimental design of this project aimed to reduce the number of variables so that materials which were effective as protection systems from literature could be compared on an equal basis in practice for their potential to inhibit corrosion of large industrial objects for at least three years. The 18 protection systems which appeared most likely to meet the project's requirements of longevity, appearance, availability, stability and safety were evaluated at equal dry film thicknesses by mechanical, atmospheric and accelerated corrosion tests. A summary of the performances of the coatings based on mechanical, exposure and oxygen consumption examinations is shown in Table 2.

Both hardness and adhesion tests appeared to be poor predictors of the ability of coatings to protect either in real time or accelerated conditions. The ability of coatings to protect against corrosion as shown by salt spray testing for 504 h and atmospheric aging in the six month period studied were identical. Based on these results, it can be concluded that salt spray tests represent real time aging accurately in this study.

Bare Q panels corroded significantly more rapidly than all coatings, suggesting that any treatment is better than nothing. Coatings which clearly inhibit corrosion, as shown by salt spray tests, atmospheric corrosion after six months and oxygen consumption are Cosmoloid H80, Dinitrol Car/4941, LPS3, Rustilo 3000, SP400 and VpCI-386. Although this is an interim result because the ideal requirement is for the coatings investigated to withstand real time corrosion for three years, the accelerated salt spray and oxygen consumption results are good indicators of the protection ability of the materials. The selected coatings meet the ideal requirements for health and safety and availability. The requirement for appearance must be considered from case to case, but the selected coatings produce both transparent and opaque finishes.

The next stage of the project is to examine reversibilities and performances of coatings on precorroded Q panels as well as on real artefacts. Most industrial objects require conservation because they are corroded. Removing corrosion layers completely is time consuming and expensive and is not always desirable if original surfaces are to be preserved. The performance of the selected protection systems on precorroded surfaces will be evaluated using the same procedures as those applied to bare Q panels.

Acknowledgements

The authors wish to thank Michel Malfilâtre for applying all coatings. Thanks are also due to Ship-2-Shore, Eurodeal Autoparts A/S and Th.C. Tromm GmbH for providing samples of their products.

References

1. Cortec Corporation, St. Paul, MN, USA, available at: http://www.cortecvci (accessed on 25 January 2010).
2. Ship-2-Shore Liquid Corrosion Control Systems, available at: http://www.ship-2-shore.com
3. Castrol Lubricants (S) Pte Ltd, CWT Distripark, Singapore, Singapore.
4. B. Seipelt, M. Pilz and J. Kiesenberg: Proc. Int. Conf. on 'Metals Conservation', Draguignan, France, May 1998, James and James (Science Publishers) Ltd, 291–296.
5. D. Hallam, D. Thurrowgood, V. Otieno-Alego, D. Creagh, A. Viduka and G. Heath: Proc. Conf. ICOM Committee for Conservation Metals Working Group, Santiago, Chile, April 2001, Western Australian Museum, 297–303.
6. V. Argyropoulos, D. Charalambous, D. Poilkreti and A. Kaminari: 'Strategies for saving our cultural heritage', Proc. Int. Conf. on 'Conservation strategies for saving indoor metallic collections', Cairo, Egypt, February–March 2007, TEI of Athens, 115–120.
7. Alvin Products, available at: http://www.hightempenginepaint.com (accessed on 25 January 2010).
8. Delphi DX900165 Thumbnail hardness test for painted parts, available at: http://auto.ihs.com/document/abstract/YDBRIAAAAAAAAAAA (accessed on 25 January 2010).
9. H. Matthiesen and K. Wonsyld: '*In situ* measurement of oxygen consumption to estimate corrosion rates', Proc. EUROCORR 2009, Nice, France, September 2009, the European Federation of Corrosion, Paper WS D-O-7856.

Evaluation of new non-toxic corrosion inhibitors for conservation of iron artefacts

S. Hollner[1], F. Mirambet*[1,2], E. Rocca[3] and S. Reguer[4]

The anticorrosion performances of new non-toxic inhibitors based on carboxylic acids extracted from vegetable oil have been evaluated for the protection of iron artefacts. Electrochemical measurements and natural aging tests have demonstrated the efficiency of those inhibitors in the context of temporary treatments. Surface analysis coupled with *in situ* X-ray absorption near edge structure experiments has revealed that their anticorrosion properties are correlated to the precipitation at the metal surface of a protective layer made of iron carboxylate.

Keywords: Atmospheric corrosion, Cultural heritage, Protection, Iron carboxylate, Corrosion inhibitor

This paper is part of a special issue on corrosion of archaeological and heritage artefacts

Introduction

Owing to temperature and humidity fluctuations, metallic elements of the cultural heritage are exposed to cyclic wet and dry periods and suffer indoor atmospheric corrosion. This process is the main cause of their decay and sometimes leads to their total degradation. Thus, the development of new protection systems, which can contribute to a decrease in the corrosion rate of metallic artefacts, is one of the main objectives of research in conservation.

For many years, conservation scientists share their effort to develop new anticorrosion systems, which involve the creation of a more or less insulating layer between the metal and its corrosive environment.[1,2] For industrial purpose, many different protection systems have been produced, such as organic coatings, ceramic and inorganic coatings, conversion coatings, corrosion inhibitors. Unfortunately, most of them are not adapted to the requirements of the cultural heritage domain mainly because they lead to important changes in the visual appearance of the object and even require the removal of the corrosion layer, which is not possible in the case of much corroded archaeological artefacts as well as patinated objects.

Among the protection systems tested by conservation scientists on real artefacts for 20 years, it appears that corrosion inhibitors and coatings, like waxes and varnish, could fulfil these requirements.[3,4] Unfortunately, most of the corrosion inhibitors commonly used by conservators for the protection of metallic artefacts like benzotriazole are hazardous to human health.

Despite the numerous corrosion inhibition investigations, relatively few works are directed towards the study of non-toxic organic compounds.

Therefore, a major advancement would be reached in the conservation domain by the development of new environment friendly inhibitors, which can be easily applied without expensive conservation works.

Within the framework of the European project 'Promet',[5] the authors developed new safe corrosion inhibitors for iron based artefacts. These compounds with the general formula $CH_3(CH_2)_nCOOH$, noted as HC_n, are generally used in the form of sodium carboxylate, noted as NaC_n. They are environmentally friendly compounds extracted from vegetable oils (sunflower, colza and palm), easy to remove and not very expensive. This family of inhibitors is known to have good inhibition properties on copper, zinc, lead, magnesium and aluminium,[6–10] and they are now studied for application on iron samples.[11] In this paper, the authors present the evaluation of the performances of NaC_{10}, HC_{10} and HC_{14} treatments on bare iron simulating iron artefact by electrochemical measurements. The protection mechanism was investigated through the characterisation of the protective layer by scanning electron microscopy, Raman spectroscopy and by *in situ* X-ray absorption near edge structure (XANES) experiments performed on a Fame BM30b beamline at the European Synchrotron Radiation Facility (Grenoble, France).

Finally, the more effective solution was tested on real artefacts belonging to the National Maritime Museum of Paris to evaluate their effectiveness in real conditions.

Experimental

Experiments have been carried out on iron coupons called 'Promet coupons' (low carbon steel, 0·14 wt-%C) produced to simulate iron based objects. These coupons have been precorroded after exposition in humid chamber

[1]LRMH, Laboratoire de Recherche des Monuments Historiques, 29, rue de Paris, F 77420 Champs-sur-Marne, France
[2]C2RMF, Centre de Recherche et de Restauration des Musées de France, Palais du Louvre, Porte des Lions 14 Quai François Mitterrand, 75001 Paris, France
[3]Institut Jean Lamour, CP2S-UMR CNRS 7198, Nancy Université, BP 70239, 54506 Vandoeuvre-lès-Nancy Cedex, France
[4]Synchrotron SOLEIL, L'Orme des Merisiers, Saint-Aubin, BP 48, 91192 Gif-sur-Yvette Cedex, France

*Corresponding author, email francois.mirambet@culture.gouv.fr

Published by Maney on behalf of the Institute
Received 11 January 2010; accepted 2 May 2010
DOI 10.1179/147842210X12732285051311

[24 h at 30°C/100% relative humidity RH+24 h at 25°C/50–60% RH+24 h at 30°C/100% RH]. The artificially aged coupons were further mechanically cleaned to reproduce the surface of real objects.

Linear sodium decanoate (noted as 'NaC_{10}') solutions (0·1 mol L^{-1}=19·4 g L^{-1} and 0·05 mol L^{-1}=9·7 g L^{-1}) were prepared through neutralisation of decanoic acid by sodium hydroxide in water. The solutions of carboxylation (noted 'HC_n solutions'), which are very similar to that of phosphatation, were prepared by mixing in a hydro-organic medium (50% water+50% ethanol) a carboxylic acid HC_n with an oxidising agent at 0·1 mol L^{-1} hydrogen peroxide (H_2O_2). The final pH of the solution is ~3·5. The following concentrations were used: 30 g L^{-1} for HC_{10} and 20 g L^{-1} for HC_{14}. In the case of HC_{14}, the solution was heated around 40–50°C to allow good solubilisation. For this last solution, a reheating ~40°C before application is needed.

The corrosive medium simulating the atmospheric corrosion was the ASTM D1384-87 standard (noted as 'ASTM water'), which has the following composition: 148 mg L^{-1} Na_2SO_4, 138 mg L^{-1} $NaHCO_3$ and 165 mg L^{-1} NaCl.

Electrochemical measurements were performed on iron Promet coupons in electrolyte solutions prepared by adding to ASTM water different amounts of NaC_{10} solution ranging from 5×10^{-2} to 10^{-1} mol L^{-1} (pH 7·5–8·5). For the carboxylation solutions, the Promet coupons were previously immersed in the HC_{10} and HC_{14} solutions (1 or 4 h of immersion), then electrochemical measurements were performed in the ASTM water. Electrochemical tests were carried out in a three-electrode electrochemical cell connected to a Gamry PCI4/300 potentiostat. The circular and horizontal working electrode (3·14 cm^2) was placed at the bottom of the cell under a Pt disc electrode. The reference electrode was a KCl saturated calomel electrode [Hg/Hg_2Cl_2, E=+0·242 V(SHE)].

The polarisation resistance R_p measurements were carried out with a scan rate of 0·125 mV s^{-1} in the range of 20 mV ($E_{corr}=\pm10$ mV).

The potentiodynamic curves i=f(E) were recorded after 8 h of immersion from −200 to 1100 mV versus E_{corr} with a sweep rate of 3 mV s^{-1}.

Solid iron decanoate powder FeC_{10} was synthesised by mixing an iron(III) nitrate aqueous solution with a NaC_{10} solution. The orange red precipitates were filtered, rinsed with distilled water and dried in an oven for a few hours at temperatures between 60 and 70°C.

Surface characterisation was carried out with a field emission gun SEM (Hitachi S4800, Hitachi, Tokyo, Japan) and by a Raman spectroscopy on a Jobin Yvon LabRam Infinity spectrometer (Jobin-Yvon Inc., Edison, NJ, USA), with an Nd/YAG laser emitting at 532 nm and coupled to an Olympus microscope allowing sample micrometric observation and analysis.

In situ X-ray absorption spectroscopy (XAS) experiments were performed on a BM30b beamline at the European Synchrotron Radiation Facility using a specific homemade three-electrode cell, completely described in a previous paper.[12] The reference electrode was a calomel one, and the counter electrode was a platinum wire. A composite electrode made of graphite powder mixed with iron powder fixed on amorphous carbon plate constitutes the working electrode. The electrolyte was 0·1 mol L^{-1} NaC_{10}+NaCl solution. The system was polarised at E=1·5 V(SCE) in order to increase the oxidation rate.

Measurements by XANES were performed at room temperature with a 200×300 μm^2 beam size at sample position. Spectra (XANES) were acquired using a double crystal Si(220) monochromator from 100 eV before to 700 eV after the edge position, with a 5 eV step size in the pre-edge region, 0·3 eV in the edge region and constant 0·5 $Å^{-1}$ k-step in the post-edge region. The measurements were collected in fluorescence mode using a 30-element Canberra detector. The calibration is obtained with Fe foil by setting the first peak of the first derivative of the spectrum at 7112 eV.

The application of the protection system was realised on a real artefact: a part of a chain of a marine anchor conserved in the storage rooms of a maritime museum located in the Romainville fortress near Paris.

Before the application of the treatment solution, a part of the surface of the chain was cleaned by application of a nitric acid solution to remove the non-adherent corrosion products. The surface aspect after cleaning is very close to those observed on Promet coupons.

Results and discussion

Electrochemical measurements

NaC_{10} solutions

Figure 1*a* shows that an increase in the R_p values with the immersion time is systematically observed for the solutions containing two different concentrations of NaC_{10}. After 7 h of immersion, the R_p values reach a 50–60 kΩ cm^2 range. The concentration effect on the performance of NaC_{10} solutions for short immersion times is clearly demonstrated since, below 4 h, the higher R_p values are recorded for the most concentrated solution (0·1M NaC_{10}).

Potentiodynamic curves recorded after 8 h of immersion are presented in Fig. 1*b*. The comparison to the curve corresponding to the inhibitor free solution shows that sodium decanoate induces a strong modification of the electrochemical behaviour. A drastic shift of the E_{corr} values from −0·8 to −0·2 V(SCE) and a decrease in the corrosion current are clearly observed, which is mainly due to the decrease in anodic reaction rate of corrosion. Moreover, the i=f(E) curves display a passivation plateau over a large potential range with a low passive current density [i_p, ~$2\cdot10^{-6}$–10^{-5} A cm^{-2} below 0·6 V(SCE)] in comparison with the values recorded for the inhibitor free ASTM water.

Carboxylation treatments

These treatment solutions were developed to allow the precipitation at the metal surface of conversion coating as in the case of phosphating process. The electrochemical tests were performed in the ASTM corrosive medium after treatment of the Promet coupons in the carboxylation solutions.

Figure 2*a* shows the corrosion potential evolution of treated Promet coupons with immersion time in the ASTM corrosive medium. For the non-treated and NaC_{10} treated coupons, the corrosion potential rapidly decreases to −0·7 V after 2 h. In comparison, the E_{corr} of the coupons treated in HC_{10} solution decreases but remains at higher values. At last, with the HC_{14} solution,

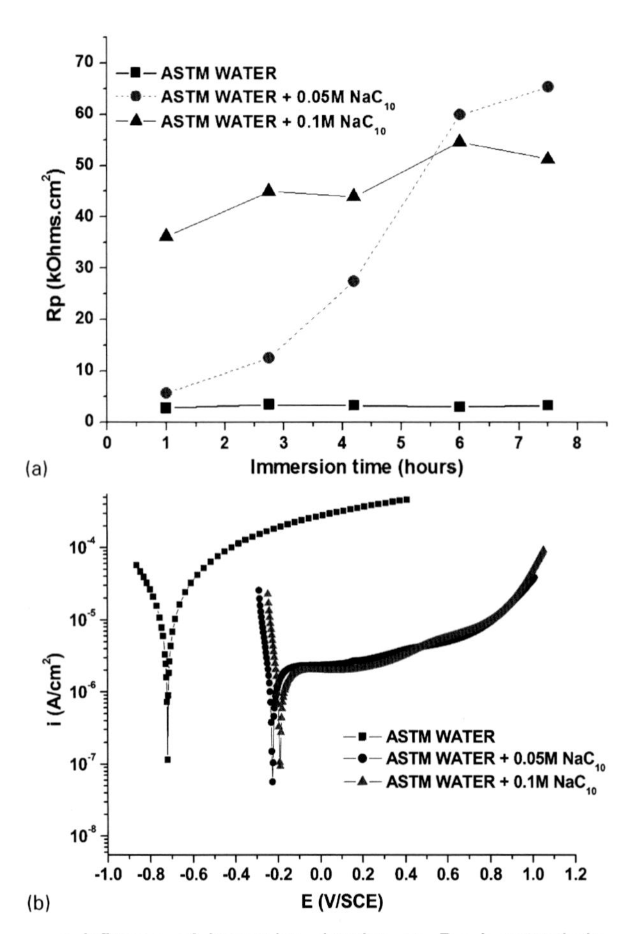

a influence of immersion duration on R_p; *b* potentiodynamic curves of iron recorded after 8 h of immersion

1 Electrochemical measurements performed on Promet coupons immersed in ASTM solution containing different amounts of NaC_{10}

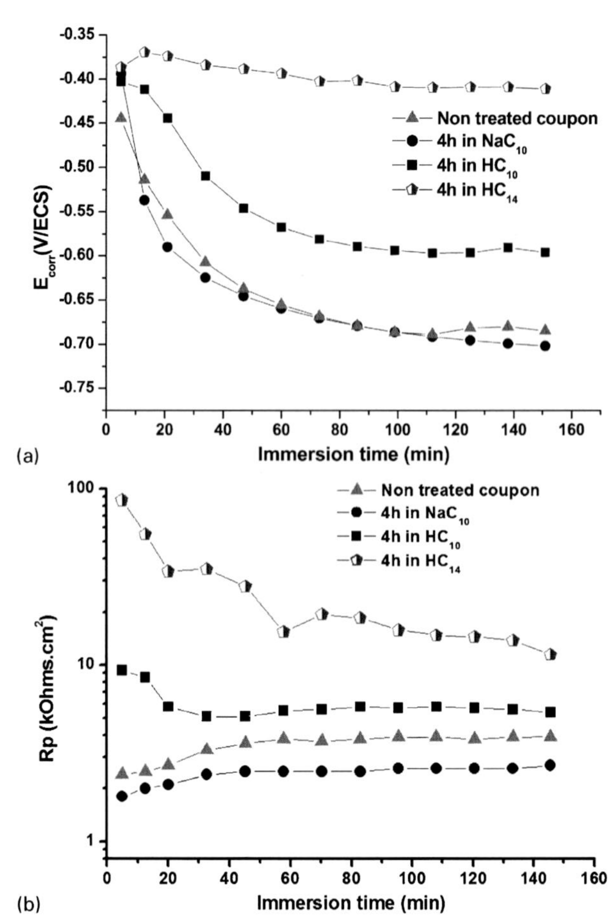

2 Influence of immersion duration on *a* corrosion potential and *b* R_p values of Promet coupons

the E_{corr} remains stable throughout the experiment. These results seem to prove the better efficiency of the carboxylation solutions in relation to the sodium decanoate ones.

The R_p values measured in the same conditions confirm the good performance of the carboxylation treatments (Fig. 2*b*). At the beginning of the immersion, the R_p values of the Promet coupons treated in the carboxylation solutions are almost one or two orders of magnitude higher than those recorded with the non-treated and NaC_{10} treated coupons. Then, a decrease in the R_p values is observed with the immersion time. These experiments show that HC_{10} and HC_{14} carboxylation treatments provide better protection than immersion in NaC_{10} solution. The protection can be easily improved by the use of carboxylic acid with longer carbon chain length like HC_{14}.

Surface characterisation

The SEM observations reveal that the surfaces of samples treated in HC_n solutions are very well covered by a passive layer made of more or less tangled crystals (Fig. 3). Moreover, the crystal size increases with the carbon chain length as observed in Fig. 3*b* and *c*. As displayed in Fig. 4, the Raman spectrum of the passive layer formed after treatment in HC_{10} solution shows several characteristic vibration bands of a synthetic iron decanoate especially around 2800–3000 cm^{-1} (CH_3 groups) and 1500 cm^{-1} (COO groups). The Raman spectrum recorded on the corroded zones of the Promet sample also reveals the presence of iron carboxylate in association with iron oxyhydroxide (goethite and lepidocrocite). This result confirms that an iron carboxylate layer has also precipitated on rust pits (Fig. 4).

On the contrary, only very weak bands around 2800–3000 cm^{-1} were detected by Raman spectroscopy on the sample treated in NaC_{10} solution. As confirmed by the SEM image of the treated sample (Fig. 3*a*), the inhibition of iron corrosion by sodium carboxylate solution is due to the formation of a very thin layer of iron carboxylate as previously observed in the case of copper.[13]

X-ray absorption spectroscopy experiments

In order to follow the passivation process and to characterise the compound formed *in situ* during the treatment, XANES measurements were performed on iron powder sample immersed in NaC_{10} solution. The system was polarised at E=1·5 V(SCE) in order to increase the oxidation rate. Before any application of an oxidation potential, the XANES spectra acquired on the working electrode corresponds to the iron metal signal with edge at 7112 eV. The XANES spectrum was recorded after 2 h of immersion. The energy shift of the edge from 7112 eV to higher energies (~3 eV) indicates an increase in the oxidation state (Fig. 5). The comparison with the spectrum of FeC_{10}, previously synthesised, confirms that the passivation mechanism of iron coupon immersed in non-toxic inhibitive solutions is correlated to the precipitation of an insoluble Fe(III) carboxylate.

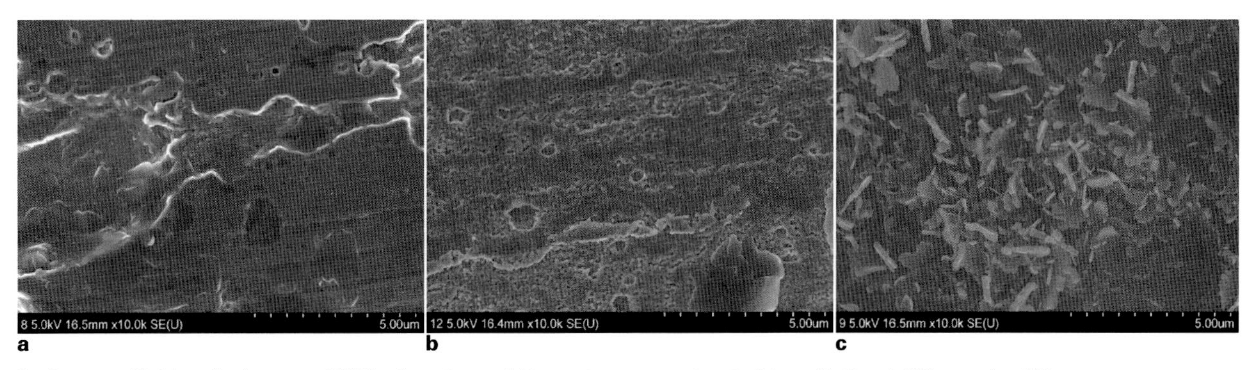

3 Images (field emission gun SEM) of surface of Promet coupons treated in *a* NaC_{10}, *b* HC_{10} and *c* HC_{14}

Natural aging tests

The final testing step was carried out on a part of a marine anchor belonging to the National Maritime Museum of Paris. This artefact is stored in a boathouse in uncontrolled environmental conditions.

For a clear evaluation of the inhibitor performance, only half of the surface previously cleaned was treated with the carboxylation solution (surface surrounded by a red square on the chain in Fig. 6).

Considering the electrochemical results and the fact that the HC_{14} solution has to be warmed up to be applied, only the HC_{10} solution, which gathered the best compromise between efficiency and easiness of application, was tested during the natural aging tests.

Control of the surface aspect was carried out each month to evaluate the efficiency of the protection system tested.

After nine months of exposure in very bad conditions, only the unprotected part of the artefact is now covered with rust pits. No modification of the surface was observed after application of the treatment solutions, and at the end of the tests, the treated surface remained uncorroded, confirming the efficiency of the HC_{10} treatment in real conditions of storage.

Discussion

This study proves that NaC_{10} and HC_n solutions are effective corrosion inhibitors for iron in a corrosive medium simulating atmospheric corrosion.

The protection properties of these treatments on bare iron are linked to the formation at the metal surface of a hydrophobic layer, already observed for other metallic surfaces[7–9] made of iron carboxylate.

Electrochemical results coupled with surface characterisation show that sodium decanoate solutions allow the formation of a very thin protective layer. On the contrary, the joint effect of the carboxylic acid and the oxidising agent certainly enhances the release of iron cations, leading to the precipitation of thicker iron carboxylate layers, which improves the anticorrosion performances.

This new family of non-toxic conversion treatment should be used as temporary treatment in sheltered conditions but can be prepared and applied very easily and safely in comparison with common corrosion inhibitor used by conservators, like benzotriazole. Moreover, it fulfils the main conditions of application laid down by conservation ethical rules. As observed on the marine anchor, their applications do not modify the surface appearance, and the iron carboxylate layers formed are easily removable with a solvent like ethanol, allowing a complete reversibility of the treatment.

These properties are of great interest in the case of the protection of large metallic collections, which cannot be conserved in uncontrolled environmental conditions and for which the numbers of pieces that have to be treated is a limiting factor for the application of common protection system like waxes or varnish. These problems are most frequently encountered in the case of scientific,

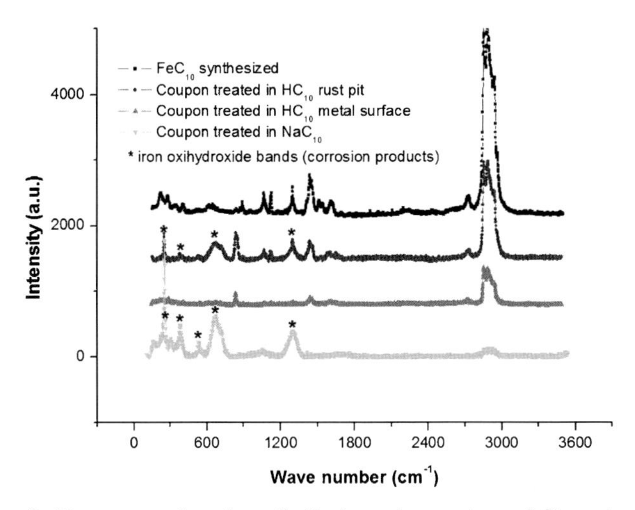

4 Raman spectra of synthetic iron decanoate and Promet coupon treated in HC_{10} and NaC_{10}

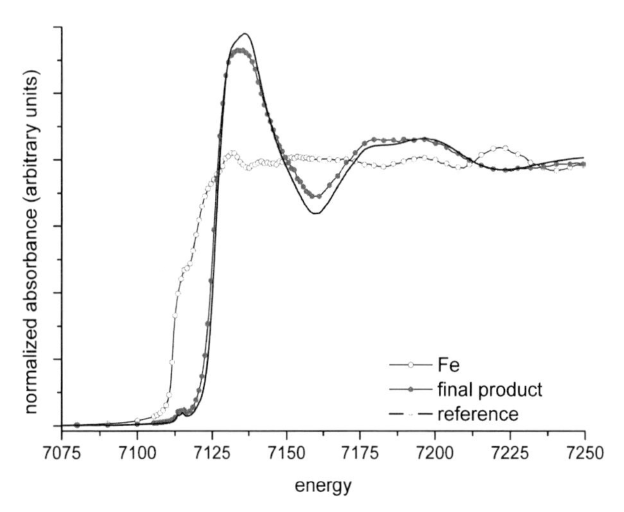

5 X-ray absorption spectroscopy spectra recorded during *in situ* experimentation on BM30b beamline: reference compound is iron(III) carboxylate FeC_{10} prepared by mixing an iron(III) nitrate aqueous solution with NaC_{10} solution

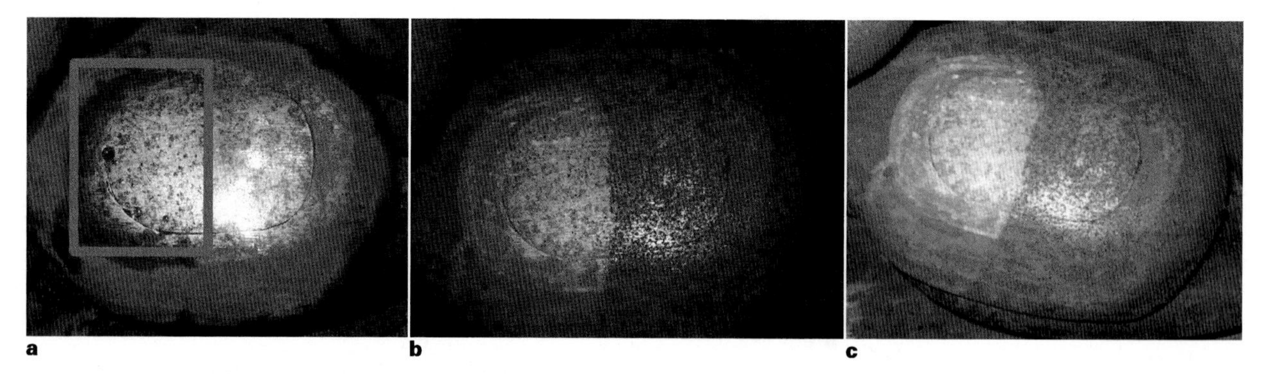

a March 2007; *b* December 2007; *c* August 2008

6 Evaluation of performance of HC_{10} solution by natural aging test: examination of marine anchor surface for different periods of test

technical and industrial collections generally stored in old buildings.

The use of these new corrosion inhibitors based on carboxylic acids with different carbon chain lengths extracted from vegetable oil is also a new opportunity for curators to reduce the corrosion rate of their metallic collections; meanwhile, they could find funds to organise a restoration campaign or to improve the environmental storage conditions.

Conclusions

For more than 40 years, many conservation scientists have carried out research programmes based on the evaluation of the performances of anticorrosion systems, which could be commonly used for the protection of metallic artefacts.

In this context, the present study confirms that the use of sodium decanoate solution and carboxylation treatments as temporary anticorrosion protection seems to be very promising.

These new temporary treatments, which can be easily prepared and applied on metallic surfaces, will permit avoiding large and expensive conservation works and could solve the problem of safe storage of large metal collections exposed to uncontrolled environmental conditions. Thus, these new corrosion inhibitors are tested on real metallic collections in close collaboration with several 'Musées of France'.

References

1. C. Price, D. Hallam, G. Heath, D. Creagh and J. Ashton: Proc. Conf. Metals'95, 223–241; 1996, London, James and James Ltd.
2. K. Rahmouni, N. Hajjaji, M. Keddam, A. Srhiri and H. Takenouti: *Electrochim. Acta*, 2007, **52**, 7519–7528.
3. R. B. Faltermeier: *Stud. Conserv.*, 1998, **43**, 121–128.
4. L. B. Brostoff: Proc. Conf. Metals'95, 99–108; 1996, London, James and James Ltd.
5. V. Argyropoulos (ed.): 'Metals and museums in the Mediterranean', 'The PROMET project'; 2008, Athens, TEI of Athens.
6. C. Rapin and P. Steinmetz: Proc. Conf. Corrosion' 98, San Diego, CA, USA, March 1998, NACE International, Paper 211.
7. D. Daloz, C. Rapin, P. Steinmetz and G. Michot: *Corrosion*, 1998, **54**, 440–444.
8. E. Rocca and J. Steinmetz: *Corros. Sci.*, 2001, **43**, 891–902.
9. E. Rocca, C. Rapin and F. Mirambet: *Corros. Sci.*, 2004, **46**, 653–665.
10. E. Rocca, G. Bertrand, C. Rapin and J. C. Labrune: *J. Electroanal. Chem.*, 2001, **503**, 133–140.
11. P. Bommersbach, C. Alemany-Dumont, J. P. Millet and B. Normand: *Electrochem. Acta*, 2005, **51**, (6), 1076–1084.
12. J. Monnier, L. Legrand, L. Bellot-Gurlet, E. Foy, S. Reguer, E. Rocca, P. Dillmann, D. Neff, F. Mirambet, S. Perrin and I Guillot: *J. Nucl. Mater.*, 2008, **379**, 105–111.
13. G. Bertrand, E. Rocca, C. Savall, C. Rapin, J.-C. Labrune and P. Steinmetz: *J. Electroanal. Chem.*, 2000, **489**, 38–45.

Use of artificial metal coupons to test new protection systems on cultural heritage objects: manufacturing and validation

C. Degrigny*

New protection systems for cultural heritage artefacts can be tested and their efficiency, compared to traditional systems, assessed using artificial metal coupons. The manufacturing of these coupons requires though a thorough investigation of the artefacts that they are supposed to simulate. Not only their composition, but their surface preparation (through artificial or natural aging) should copy as closely as possible real artefacts. Protection systems tested should be applied according to standardised protocols, while short term testing in humid chamber and long term testing on site should be monitored regularly to detect any possible failure. Only then the use of such artificial metal coupons will be validated. This approach is illustrated in this paper through the EU PROMET and Swiss POINT research projects.

Keywords: Artificial coupons, Steel, Brass, Protection systems, Testing, Conservation

This paper is part of a special issue on corrosion of archaeological and heritage artefacts

Introduction

When artificial metal coupons are mentioned in the conservation–restoration (C–R) literature, it is often to expose them on site to appreciate the corrosivity of a storage environment or an exhibition area.[1–3] Indeed, metals are sensitive to the action of certain pollutants: Ag tarnishes in sulphur atmospheres (mainly H_2S),[4] Cu and particularly Pb are corroding in the presence of formic and acetic acid vapours.[5–7] Since the corrosion level depends on the concentration of pollutants in the atmosphere, these metals have naturally been considered to develop sensors used to determine and monitor the amount of pollution in heritage sites where archaeological and historic metals are conserved. The EU funded Corrlog project is a good illustration of the possibilities of such sensors.[3] The 'Oddy' test is another application of these measuring systems. Clean Ag, Cu and Pb coupons, representing the surface of heritage objects, are, in this test, inserted separately in glass containers with the testing material that is supposed to produce corrosive pollutants and distilled water (in a separate small container) to maintain a 100% relative humidity (RH). To accelerate the corrosion processes, the containers are placed during 4 weeks in an oven at 60°C. Alterations are documented at the end of the testing period.[8]

The drawback of these sensors is that the metal surface exposed does not correspond, except in the case of Ag based alloys traditionally exhibited in their polished condition, to the corroded aspect of archaeological and historic objects. Cu based objects are normally covered with an artificial or natural patina, appreciated aesthetically but the composition of which is different from that of slightly oxidised metals.[9] The corrosion layers formed on Pb based alloys, while being protective, give a tarnished appearance.[5] Without knowing the reactivity of these corrosion layers to atmospheric pollutants, it is difficult to conclude on the real reactivity of heritage objects to the same pollutants. Therefore, these sensors are mainly used in preventive conservation strategies, as early warning systems.

If the improvement of storage and exhibition conditions is a good way to slow down, or even stop, any future alteration of a metal object, there is another option, which is to protect its surface. Once again, a lot of scientific work has been carried out in the past to test different protection systems on clean metal coupons.[10–12] The first time that artificial metal coupons simulating the alteration of historic objects were used was in order to study the protection of outdoor bronzes and industrial steel structures in uncontrolled atmospheres. Testing materials for bronzes were coupons exposed on specific sites.[13–15] These coupons have obviously limited use since the corrosion layers are different from one site to another. As regards materials simulating industrial structures, steel sheets were precorroded artificially.[16] Cu based coupons were only degreased before the application of the protection systems tested, while thick corrosion layers present on steel coupons had to be cleaned mechanically to recover the original surface.

The author's literature review showed that while authors are convinced of the necessity to use artificial coupons to test/compare the efficiency of protection systems, the following questions are not well documented:

(i) the choice of the base material that plays a major role in the development of specific

Haute Ecole de Conservation–Restauration Arc, 60 rue de la Paix, CH-2301 La Chaux-de-Fonds, Switzerland

*Corresponding author, email christian.degrigny@gmail.com

Published by Maney on behalf of the Institute
Received 16 January 2010; accepted 29 May 2010
DOI 10.1179/147842210X12754747500649

corrosion forms and products.[6] Often, the materials used to manufacture metal coupons are easily available in the market and are considered to be more or less representative of the studied materials: bronzes, 85Cu–5Sn–5Zn–5Pb (wt-%) or 90Cu–5Sn (wt-%); brass, 65Cu–35Zn (wt-%) or Cu for Cu based alloys and low carbon steel for iron based alloys

(ii) the aging processes. Usually, artificial aging is favoured to natural aging because it is quick, and experts consider that it is almost impossible to obtain, in a reasonable time, historic corrosion layers with a natural aging process

(iii) the cleaning of coupons before the application of the protection systems. A well cleaned surface is normally required for good efficiency of the protection. The approach is specific in C–R since the C–Rs try to conserve the original surface of metals within the corrosion layer while removing the powdery corrosion products

(iv) the representativeness of corrosion layers finally obtained. Protection systems will not be similarly efficient on a metal surface covered with just one adherent and homogeneous corrosion product (often obtained during artificial aging) and a heterogeneous mixture of poorly adherent corrosion products (that are closer to the real situation).

Data concerning the application and the monitoring of the protection systems tested on artificial metal coupons are not well documented either.

Without this information, it is hard to validate the use of artificial metal coupons to assess the efficiency of protection systems on heritage objects. In the following and through two examples, the author will try to confirm the real interest of considering this preliminary step before applying protection systems on real artefacts. The first example has been taken from the EU funded PROMET project (2004–2008) that aimed at studying and protecting metal objects from museums in the Mediterranean basin with safe and innovating protection systems.[17] In that work, the tested protection systems were first compared on artificial coupons and finally assessed on real artefacts. The second is from the Swiss POINT project, which is currently ongoing and is a continuation of the PROMET project. Its objective is to apply some of the best rated protection systems developed in the PROMET project on a selection of objects of the collections of the Swiss Army Historical Material Foundation (HAM Foundation) and assess their efficiency.[18]

Criteria to prepare artificial metal coupons

The questions mentioned above are addressed in the following.

Choice of base material

The first duty of any C–R professional that intends to apply adequate protection systems on a certain category of objects (outdoor bronzes or others) or a collection of objects (armour elements from a military museum or others) should be to carry out a condition survey on these objects to determine the families of materials encountered, their condition state and the risks of alteration with time. It is through such condition surveys performed on their collections that the partners of the EU PROMET project chose a low carbon steel (0·14 wt-%C) to represent the archaeological and historic iron based artefacts examined.[19] Cold working was the main manufacturing technique observed on these objects, but from antiquity to classical periods, the processes have changed. Since the most important collection of iron based objects came from the Palace Armoury, Valletta (Malta) and a thorough destructive investigation of a representative selection of artefacts showed that most of them were constituted of a Widmanstätten microstructure (with numerous non-metallic slag inclusions), this one was considered for the manufacturing of the artificial coupons.[20] The steel sheets were heat treated (4·5 h at 950°C) and cooled rapidly to obtain this microstructure. The metal contained only a very small amount of non-metallic slag inclusions. The PROMET Cu based coupons were manufactured from a quaternary bronze alloy (85Cu–5Sn–5Zn–5Pb, wt-%) commonly found in Mediterranean cast archaeological collections. Both coupons were grinded and polished to a level similar to real artefacts (silicon carbide paper, grade 4000).

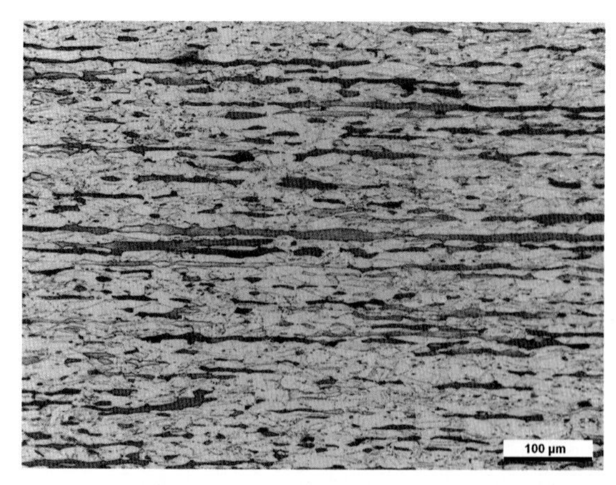

1 POINT leaded brass observed under metallographic microscope: light phase is Cu rich; dark (yellow brown) phase is Zn rich; Etched in ferric chloride (photomicrograph by S. Ramseyer)

The same approach was followed within the POINT project: after the condition survey of the collections of the HAM Foundation, a low carbon steel (0·1 wt-%C and 1 wt-%Mn) was considered as representative of iron based objects and a leaded brass CuZn39Pb as representative of Cu based objects. The coupons were purchased from the supplier as sheets in the following condition: heat laminated and cold cut. The examination of the materials on cross-sections showed that the leaded brass had conserved the trace of its lamination process, while the low carbon steel had a new grain structure (Figs. 1 and 2). For interpretation to colours in the latter figure, the reader is referred to the web version of this paper. These microstructures have naturally an important influence on the corrosion developed on these materials.

The dimensions of the PROMET and POINT coupons (50 × 75 mm) were large enough to limit any edge effects on the corrosion processes, and their thickness was comprised between 2 and 2·5 mm, except for bronze PROMET coupons that were 5–6 mm thick.

Aging processes

The quicker way to form corrosion layers on metal surfaces consists in immersing them in corrosive

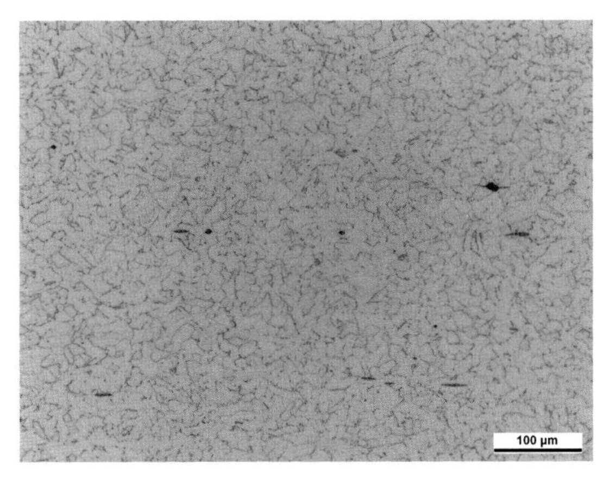

2 POINT low carbon steel observed under metallographic microscope showing MnS inclusions: etched in nital [5% (v/v) HNO_3 in alcohol] (photomicrographby S. Ramseyer)

solutions and/or exposing them in humid chambers. The major drawback of these artificial aging processes is that the layers formed do not reflect perfectly the reality: the surface appearance and surface composition might be different from those of heritage objects.

As an example, some of the PROMET steel coupons were aged artificially in a humid chamber (24 h at 30°C/100% RH+24 h at 25°C/50–60% RH+24 h at 30°C/100% RH, coupon inclination of 30°), while some of the bronze coupons where first exposed to a prechemical treatment before being aged artificially in a humid chamber (8 h at 40°C/100% RH/SO_2+16 h at 20°C/50% RH, coupon inclination of 30°). The weathering conditions were chosen according to the measurement carried out on site. Similarly, the steel/brass POINT coupons were aged artificially in a humid chamber. They were placed vertically and exposed to quicker humid/dry cycles than in the PROMET project (16 h at 40°C/100% RH+8 h at 20°C/60% RH) during 131 h for steel and 62 h for brass to accelerate the corrosion processes. If, in the case of PROMET steel coupons, the localised corrosion developed was homogeneous and reflected quite well the surface appearance of steel armour elements exposed at the Palace Armoury, Valletta (Malta), a more localised corrosion on POINT steel coupons was observed (Fig. 3*a*). This could be due to different weathering conditions (time of wetness/drying time ratio) and composition of the steel coupons in the two projects as expressed by Schmitter and Böhni.[21] While reflecting less reality of artefacts in this study, this surface appearance corresponded still to some of the slightly corroded steel artefacts of the HAM Foundation collection. POINT brass coupons were heterogeneously corroded, in a similar way as real artefacts from technical collections (Fig. 3*b*; for interpretation to colours in this figure, the reader is referred to the web version of this paper), while PROMET bronze coupons were covered with a green patina similar to the one found on outdoor bronzes.

The natural aging of metal coupons is by far the process preferred since it reflects more real alteration of heritage objects. The site of exposure is essential: indeed, a low corrosive site will lead to a very long exposure period. Within the PROMET project, the Palace Armoury, Valletta (Malta) appeared as a very well chosen exposure site. Indeed, its corrosivity was such that, after a few weeks, the steel coupons exposed on a rack developed a well distributed localised corrosion that transformed slowly in general corrosion in a year.

Surface cleaning

In the past, the corroded surfaces of heritage metal objects were often cleaned drastically (complete stripping of corrosion products) before the application of any protection systems. Today, C–Rs make a point in conserving the original surface within the corrosion layers. PROMET and POINT steel coupons were cleaned that way with cotton swabs, moistened with ethanol and rolled lightly over the surface to collect the powdery and non-adherent orange corrosion products (Fig. 4*a*). A rotating drill equipped with a fibre brush was used afterwards with no pressure to remove any remaining powdery compounds (Fig. 4*b*). Cu based coupons did not require final mechanical cleaning since the corrosion layer formed was quite adherent. Only degreasing with acetone was necessary.

Representativeness of corrosion layers

The visual observation of artificial metal coupons should be used only as a qualitative means to appreciate the corrosion level: close or far from the appearance of real artefacts. A thorough examination of the corrosion layers is necessary to ensure that the surface of the coupons is physically and chemically similar to the surface of real artefacts.

PROMET coupons aged artificially in humid chambers and naturally on-site were analysed using X-ray diffraction and micro-Raman spectroscopy for precise composition of corrosion products after their final preparation (cleaning). These analyses revealed the presence of ferrihydrite/magnetite on steel coupons aged artificially[19] and ferrihydrite/goethite/lepidocrocite, a mixture closer to what is actually found on real artefacts,[22] on naturally aged coupons. For bronze coupons aged artificially, their surface was covered with brochantite/antlerite, while on

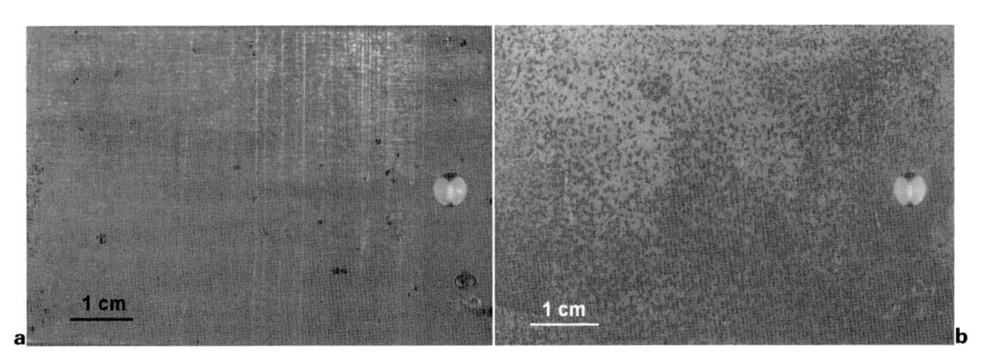

3 Surface appearance of POINT *a* steel and *b* brass coupons aged artificially (macrophotographs by A. Jaggi)

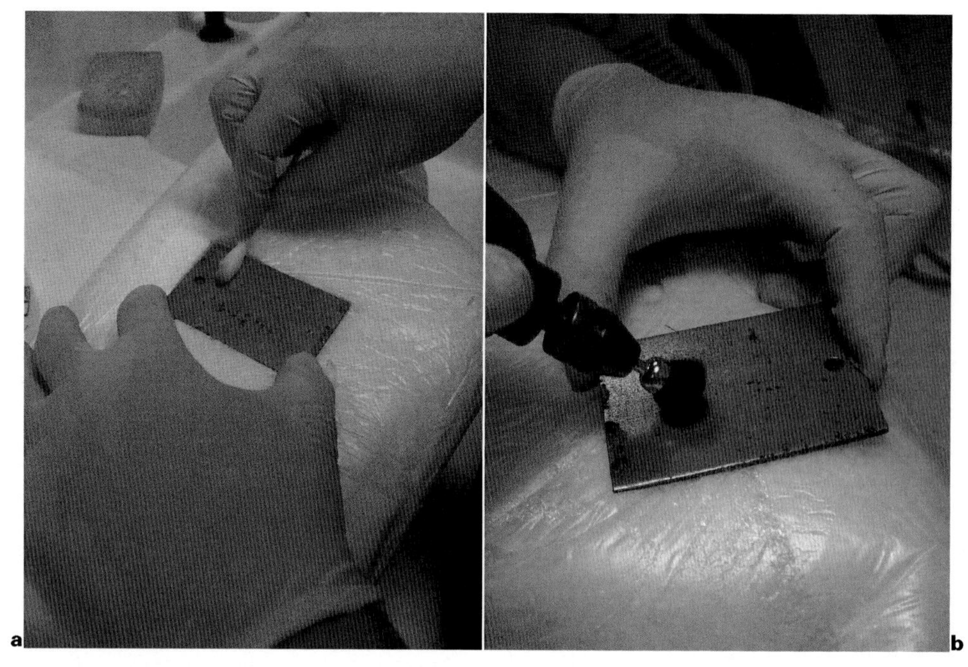

a cotton swab moistened with ethanol to collect powdery corrosion products; *b* use of rotating drill equipped with fibre brush to remove any remaining powdery compounds

4 Final preparation (cleaning) of POINT steel coupons after artificial aging in humid chamber

coupons naturally aged, only Cu oxides, hydrates and carbonates could be detected.[19]

Corrosion products that developed on POINT coupons were not analysed due to the low amount at the author's disposal.

Assessment of efficiency of protection systems on artificial metal coupons

Once artificial coupons are prepared, they can be used to test/compare the efficiency of protection systems proposed for real objects. There are a few ways to assess this efficiency (electrochemical testing, short and long term testing, etc.).[19] The ones presented as follows are the short and long term testing. These can easily be set up by C–R professionals.

Application of protection systems

The protection systems must be applied according to a defined protocol:

(i) recommendations given by the supplier if it is a commercial product or by the researcher if it has been developed within a research project

(ii) additional recommendations given by an experimented end user, for example, a C–R that is commonly applying a protection system in a specific context.

The same end user must apply the same product on all coupons tested in order to limit any risk of heterogeneity between the coupons tested.

If C–Rs often apply protection systems by brush, this technique might bring variations due to human factor. Therefore, when protection systems are first tested or their efficiency compared, the application by immersion is favoured to once again limit any risk of heterogeneity between the coupons tested. The efficiency of Renaissance microcrystalline wax, Paraloid B72 [15% (w/v) in acetone] acrylic varnish and Rylard polyurethane varnish applied by immersion (5 min) was assessed within the PROMET project on steel coupons

Table 1 Application conditions of protection systems tested by partner Heritage Malta within PROMET project on naturally aged steel coupons[19]*

Protection systems	Application	Drying
Paraloid B72 [15% (w/v) in acetone] acrylic varnish	Application by brush, one criss and one cross	Drying time: 3 h in between and 24 h at the end
Poligen ES 91009 polyethylene wax used as supplied		
Carboxylatation solution Decanoic acid HC_{10} (30 g L^{-1}) in distilled water + ethanol 50% + oxidising agent H_2O_2 (0·1 mol L^{-1})	Immersion (3 h)	Drying time: 24 h in between and at the end.
Phase vapour deposition PVD coating	Source material, Ti_3O_5; O_2 partial pressure, 4×10^{-4} mbar; total pressure, 7×10^{-4} mbar; EB gun power, 200–340 mA; coating time, 17 min	Chamber allowed to cool down to room temperature overnight and pumped up to atmospheric conditions with nitrogen gas

*Wax and varnish were applied by C–Rs. Phase vapour deposition coating and the carboxylatation solution were applied by the laboratories that developed them.

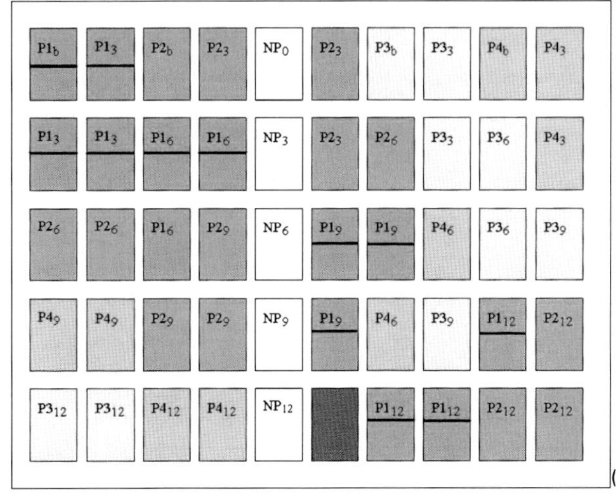

5 General view of the Perspex exposure rack used for the long-term testing of POINT coupons. *a* the vertical plate is attached to the wall and the inclined plate (30°) carries the coupons. A space is left behind the coupons so that air circulation is possible. The protected and non-protected coupons are distributed in an homogeneous way on the rack. *b* P: protection system, NP: non-protected coupon and nos. 1–4: the 4P considered. Protection systems tested on steel coupons: sodium carboxylate (NaC_{10}: one-half by immersion and one-half by brush), P1 (in blue); carboxylatation solution ($HC_{10}+H_2O_2$ by brush), P2 (in pink) and traditional protection systems, P3 and P4 (in green and yellow respectively). In index: b, white; 3, 3 months; 6, 6 months; 9, 9 months; and 12, 12 months of exposure. In red: position available

precorroded in a humid chamber (see conditions in the section on 'Aging process') and preheated at 50°C during 1 h (to dry fully the coupons and favour the application process). The number of layers applied must be put in relation to the experience of the end users, but if one system is compared to another, it is important to take into account that the protection offered should be similar: for example, one comparing the efficiency of a corrosion inhibitor developed for a short term protection to the one of a coating developed for a long term protection would not expect the same results.

The next step consists in testing the best rated protection systems in 'real' conditions. The best rated PROMET protection systems, on artificially aged coupons during the short term testing in humid chambers, were further tested on naturally aged coupons during long term testing (see below). Table 1 presents the protection systems considered by partner Heritage Malta, with application and drying conditions that are closer to reality. The efficiency of carboxylatation solution (corrosion inhibitor) applied by immersion (recommendation from the researcher), Poligen ES 91009© polyethylene wax applied by brush (recommendation from the supplier) and phase vapour deposition coating was compared to that of a reference system: Paraloid B72© [15% (w/v) in acetone] acrylic varnish applied by brush (recommendation from end users).

Testing efficiency of protection systems

The short term accelerating corrosion testing is considered as a preliminary testing to make a first choice of protection systems among a large number under study. The long term natural corrosion testing is preferred when a better idea of the real behaviour of the best rated protection systems (from the accelerating corrosion testing) on real objects is requested. If accelerating corrosion testing is commonly carried out on a short period (1 month in general), the natural corrosion testing should be conducted during at least one year or even more when the exposure site is not an aggressive one.

Short term testing

In PROMET and POINT projects, the existing ISO or ASTM standards (ISO 7384 and EN 60 068) for testing protection systems have been modified to fit the precorroded metal surfaces exposed in an indoor museum environment. In PROMET, the exposure conditions in a humid chamber of the protected coupons were taken from the aforementioned standards, reflecting extreme conditions for museum objects: 16 h at 35°C/90% RH+8 h at 20–25°C/50–60% RH, inclination of coupons different from one partner to another, during a minimum of 30 cycles. Short term testing was applied in the POINT project to optimise the application conditions of the protection systems selected. The conditions were as follows: 16 h at 40°C/90% RH+8 h at 20°C/60% RH, coupons exposed vertically, during 18 cycles.

Long term testing

Construction of an exposure rack is required. Its design reflects the collections' conditions of exposure: a rack with a slope of 30° carries coupons simulating objects freestanding in exhibition halls and suffering from dust deposition. Figure 5*a* shows the rack used in the POINT project, and Fig. 5*b* shows the distribution of the 50 tested coupons (steel or brass). With such a rack, four protection systems can be studied in parallel to reference coupons (non-protected). In addition to white coupons (protected and non-protected, non-exposed and conserved in dry conditions), groups of two to three coupons (to assess the reproducibility of the results) are exposed during periods of 3 months to 1 year (Fig. 5*b*). For interpretation to colours in the figure, the reader is referred to the web version of this paper. They are distributed on the whole surface of the rack.

Monitoring

Documentation of coupons during the short and long term testing is essential to assess the efficiency of the protection systems. Visual observation (using raking

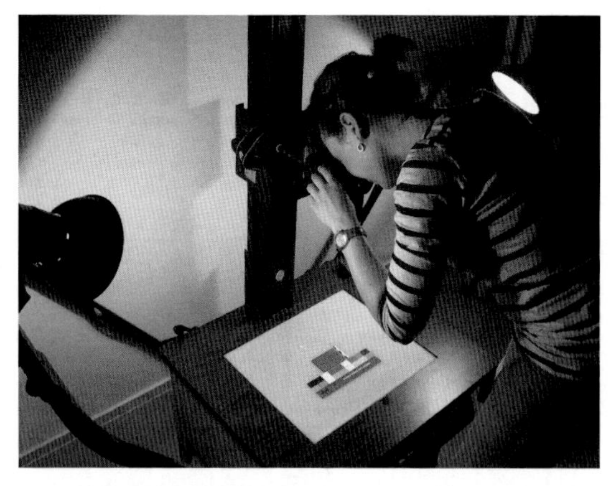

6 Standard macrophotographic documentation equipment in POINT project

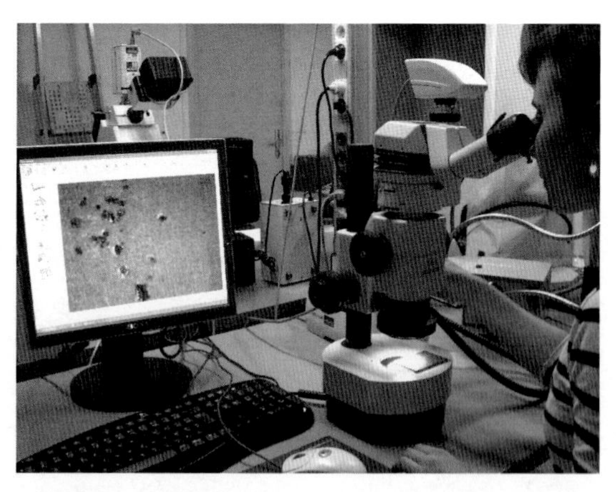

7 Standard microphotographic documentation equipment in POINT project

light) is a simple way, but any change at a microlevel is hardly detectable. Therefore, additional examination tools were considered in PROMET and POINT projects using visual, macro- as well as microscopic observations and photography with standard reproducible studio conditions, namely, same equipment, same settings and full lighting control (no variation from natural light). Two types of photographs were taken:

(i) macrophotography (Fig. 6). Each coupon was photographed in the POINT project using a Nikon D70 digital camera (at recorded settings), vertical camera stand, colour calibration card, 5 cm scales, two studio lamps with bulbs of known brand and same wattage, computer laptop and coupon stand. The coupons were handled with gloves throughout the photography session

(ii) microphotography (Fig. 7). A specific area of the coupon, representative of the whole surface or showing a default of the metal surface or the applied protection system, was chosen in POINT project with a Leica Wild M10 microscope, observed at magnification ×8 and photographed with a Leica DFC 320 camera. Data were processed with acquisition and image treatment Archimed software. Once again, the coupons were handled with gloves.

Although this was a time consuming task, such monitoring made possible a more precise observation of the way the protection system tested failed. In PROMET, Renaissance© microcrystalline wax and Paraloid B72© [15% (w/v) in acetone] acrylic varnish applied on artificially aged steel coupons seemed efficient at the beginning of the short term testing in humid chamber when observed with naked eyes, but the monitoring with microphotographs led the author to finally discard them (Fig. 8). Indeed, specific localised forms of corrosion developed after only 2 weeks (for interpretation to colours in Fig. 8, the reader is referred to the web version of this paper).

Macrophotographs were mostly used to document the changes observed in PROMET protection systems exposed to long term testing. The same photographs combined with microphotographs were particularly interesting in the POINT project, during the short term testing, in order to define the best conditions for application of the protection systems on artificially aged coupons. The way corrosion develops (localised corrosion) on non-protected artificially aged steel coupons can be clearly seen in Fig. 9a. Figure 9b shows that sodium carboxylate (NaC10) applied by brush in three layers is rather efficient since no corrosion develops, while Fig. 9c indicates that the carboxylatation solution (HC10+H2O2) applied also by brush in three layers gets damaged and favours the development of localised corrosion (Figure 9c). Therefore it is less effective (for interpretation to colours, the reader is referred to the web version of this paper). These results were expected (although not fully understood) on slightly oxidised steel surfaces

This photographic campaign requires the sampling of the coupons during the testing period to work in standardised conditions. During PROMET, an innovative standardised remote monitoring photographic

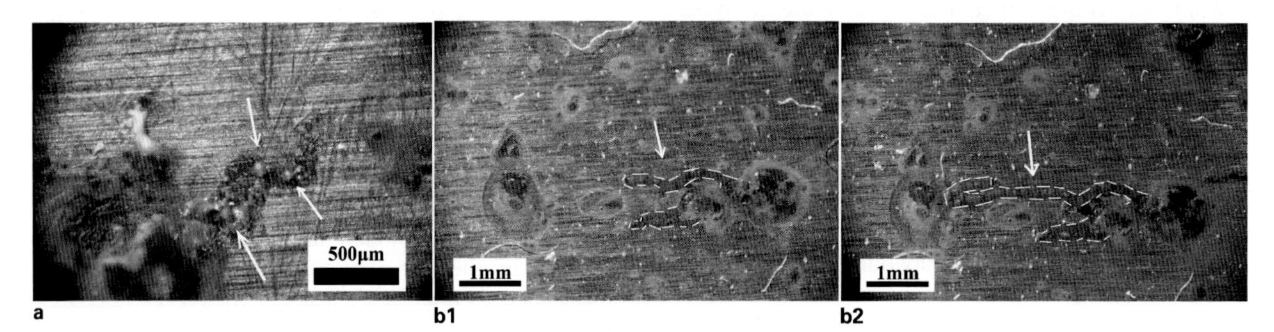

a **b1** **b2**

a local damage of Renaissance microcrystalline wax on precorroded areas after 14 cycles; *b* development of filiform corrosion under Paraloid B72 [15% (w/v) in acetone] on precorroded areas after *b*1 14 and *b*2 22 cycles

8 Microphotographs showing failure of protection systems during short term testing on PROMET steel coupons artificially aged (microphotographs by J. Crawford)

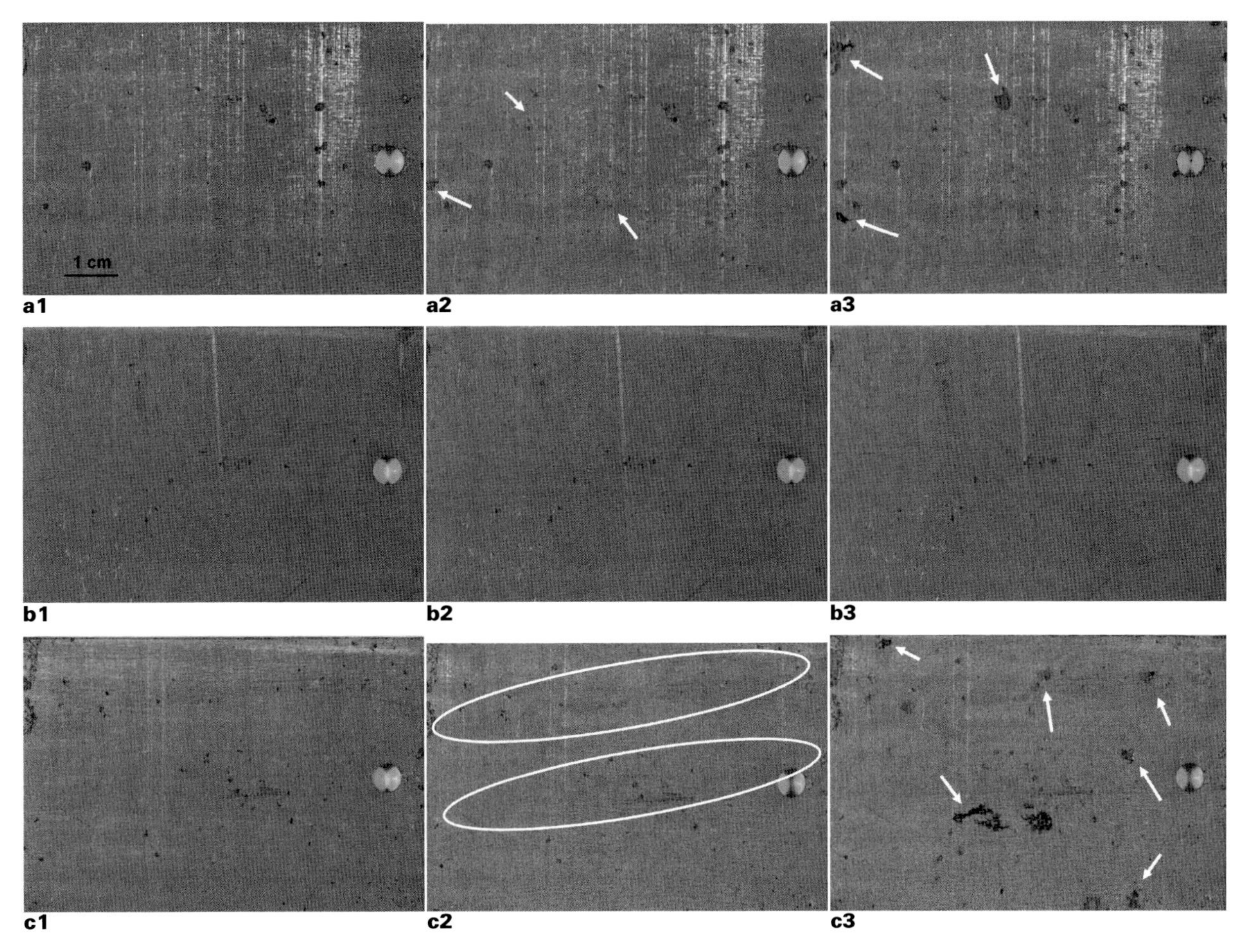

*a*1,*b*1,*c*1 for 0 day; *a*2,*b*2,*c*2 for 6 days; *a*3,*b*3,*c*3 for 12 days

9 Monitoring with macrophotographs of protected and non-protected artificially aged steel coupons during short-term testing: non-protected (a); protected with NaC10 applied by brush (three layers) (b) and protected with HC10zH2O2 applied by brush (3 layers) (c). Exposure period: 0, 6 and 12 days in humid chamber (see the protocol in section on 'Short term testing') (macrophotographs by A. Jaggi)

capture system was developed for the *in situ* documentation of the protection systems in between the examination periods.[24] A computer connected high resolution digital camera was programmed to automatically acquire reproducibly illuminated macroscopic images of the exposure rack holding the 50 coupons. The images were then analysed (digitally) to monitor changes attributable to the damage of the protection systems.

In addition to this optical documentation of the coupons that can only detect the failure of the protection systems applied and follow its development, an analytical campaign can be set up to understand the mechanisms provoking the damage observed. This campaign is easy to carry out since the coupons are transportable (small dimensions) and can be examined in non-invasive, invasive and/or destructive ways.

Conclusions

The use of artificial metal coupons to test new protection systems or compare their efficiency with that of traditional protection systems in order to protect heritage objects is of great interest. To improve the quality of the results obtained, metal coupons should be prepared to simulate as closely as possible the surface of real artefacts, the application protocols of the protection systems should be well defined and the monitoring of the protected coupons should be carried out as thoroughly as possible. The initial documentation of the whole group of coupons tested might appear as a time consuming task, but as the testing proceeds, this documentation is reduced due to the periodic removal of exposed coupons.

In the PROMET project, this methodology was employed to discard the Paraloid B72© acrylic varnish, used worldwide in the conservation field, for the protection of corroded metal surfaces exposed in uncontrolled atmospheres. Indeed, the risk of filiform corrosion developing under the film is too high. Similarly, microcrystalline waxes offered poor protection in such aggressive environments. The good behaviour of innovating and non-toxic protection systems could be validated though. Sodium carboxylates that are corrosion inhibitors offered a short term (temporary) protection, while Poligen ES 91009© polyethylene wax gave promising results for long term protection. These results are currently under validation within the POINT project.[18]

Acknowledgements

The author is grateful to the European Commission through its 6th Framework Programme, priority INCO, for the funding of the PROMET project and the Haute

Ecole Spécialisée de Suisse Occidentale (HES-SO, The University of Applied Sciences of Western Switzerland, Geneva, Switzerland) for its financial support to the POINT project as well as the research team at the Haute école de Conservation–restauration (HECR) Arc (La Chaux-de-Fonds, Switzerland) for its administrative and scientific support and the Laboratoire de Recherche des Monuments Historiques (LRMH, Champs-sur-Marne, France) for access to its research facilities.

References

1. V. Costa and M. Dubus: in 'Museum microclimates', (ed. T. Padfield, *et al.*), 63–65; 2007, Copenhagen, Nationalmuseet.
2. D. Knotkova, K. Kreislova, B. Kreibichova and I. Kudlacek: Proc. Conf on 'Strategies for saving our cultural heritage', Cairo, Egypt, February–March 2007, TEI, 64–71.
3. M. Kouril, T. Prosek, D. Thierry, Y. Degres, V. Blazek, L. R. Hilbert and M. Ø. Hansen: Proc. Metal 2007: Interim meeting of the ICOM-CC Metal WG, Amsterdam, The Netherlands, September 2007, Rijksmuseum, 78–82.
4. V. Costa: *Rev. Conserv.*, 2001, **2**, 18–34.
5. V. Costa: *Rev. Conserv.*, 2005, **6**, 48–62.
6. L. Selwyn: 'Metals and corrosion: a handbook for the conservation professional', 1st edn, 207; 2004, Ottawa, Canadian Conservation Institute.
7. C. Degrigny, E. Guilminot and R. Le Gall: Proc. ICOM-CC 11th Triennial Meet., Edinburgh, UK, Vol. 2, 865–869; 1996, London, James and James.
8. L. Green and D. Thickett: in 'Conservation science in the UK', (ed. N. H. Tennent), 111–116; 1993, London, Earthscan Ltd.
9. D. A. Scott: 'Copper and bronze in art: corrosion, colorants, conservation', 1st edn, 196; 2002, Santa Monica, CA, J. P. Getty Trust Publications.
10. E. Rocca and F. Mirambet: in 'Corrosion of metallic heritage artefacts', (ed. P. Dillmann, *et al.*), 308–334; 2007, Cambridge, Woodhead Publishing in Materials.
11. R. Combarieu, G. Dauchot and F. Delamare: Proc. Metal '98: Interim meeting of the ICOM-CC Metal WG, Draguignan, France, May 1998, James & James, 223–228.
12. V. Otieno-Alego, G. Heath, D. Hallam and D. Creagh: Proc. Metal '98: Interim meeting of the ICOM-CC Metal WG, Draguignan, France, May 1998, James & James, 309–314.
13. C. Price, D. Hallam, G. Heath, D. Creagh and J. Ashton: Proc. Metal '95: Interim meeting of the ICOM-CC Metal WG, Semur en Auxois, France, September 1995, James & James, 233–241.
14. L. B. Brostoff and E. R. de la Rie: Proc. Metal '95: Interim meeting of the ICOM-CC Metal WG, Semur en Auxois, France, September 1995, James & James, 242–244.
15. M. Pilz and H. Römich: Proc. 'Metal '95': Interim meeting of the ICOM-CC Metal WG, Semur en Auxois, France, September 1995, James & James, 245–250.
16. B. Seipelt, M. Pilz and J. Kiesenberg: Proc. Metal '98: Interim meeting of the ICOM-CC Metal WG, Draguignan, France, May 1998, James & James, 291–296.
17. V. Argyropoulos (ed.): 'Metals and museums in the Mediterranean – protecting, preserving and interpreting', 1st edn, 260; 2008, Athens, TEI.
18. C. Degrigny, F. Mirambet, A. Tarchini, S. Ramseyer, G. Rapp and A. Jaggi: Proc. Conf. on 'Conservation préventive – pratique dans le domaine du patrimoine bâti', 121–126; 2009, Bern, SCR.
19. C. Degrigny: in 'Metals and museums in the Mediterranean – protecting, preserving and interpreting', (ed. V. Argyropoulos), 179–235; 2008, Athens, TEI.
20. C. Degrigny, M. Grech, A. Williams and D. Vella: Proc. Metal 2004: Interim meeting of the ICOM-CC Metal WG, Canberra, Australia, October 2004, National Museum of Australia, 215–233.
21. H. Schmitter and H. Bohni: *J. Electrochem. Soc.*, 1980, **127**, 15–20.
22. C. Degrigny, D. Vella, S. Golfomitsou and J. Crawford: Proc. Conf. on 'Strategies for saving our cultural heritage', Cairo, Egypt, February–March 2007, TEI, 31–39.
23. S. Hollner: 'Développement de nouveaux traitements de protection à base d'acide carboxylique pour la protection d'objets en fer du patrimoine culturel', thèse de docteur, Université Henri Poincaré-Nancy I, Nancy, France, 2009.
24. J. Crawford, C. Degrigny, Q. Glorieux, P. Bugeja and D. Vella: Proc. Conf. on 'Strategies for saving our cultural heritage', Cairo, EG, February–March 2007, TEI, 85–92.

Corrosion of iron from heritage buildings: proposal for degradation indexes based on rust layer composition and electrochemical reactivity

J. Monnier*[1], P. Dillmann[2], L. Legrand[3] and I. Guillot[1]

In the present work, the authors tried to establish degradation indices for heritage ferrous artefacts, especially those used in ancient buildings and submitted to indoor atmospheric corrosion. The authors focused on the site of the Amiens Cathedral in the north of France. Samples coming from this reference site were carefully characterised in order to identify the different phases constituting the corrosion scale. The scale consists in a matrix of iron oxyhydroxide goethite embedded with several ferrihydrite marblings. Other phases such as lepidocrocite and akaganeite are scarcely present in the external part of the corrosion scale. Moreover, electrochemical measurements on both references and ancient samples enable to define the reduction reactive phases. From the nature of these phases and their localisation, two degradation indices were defined to evaluate rust reactivity. Finally, a curve that links these two factors is proposed as a first step towards a corrosion diagnosis.

Keywords: Iron, Indoor atmospheric corrosion, Heritage, Microscale analysis, Electrochemistry, Degradation diagnosis

This paper is part of a special issue on corrosion of archaeological and heritage artefacts

Introduction

Numerous ancient monuments contain high quantities of iron reinforcement in their structure. These rods and clamps have been put in the building during their construction. Examples can be found in the Beauvais Cathedral (the twelfth century) or later in the Amiens Cathedral (the fifteenth century). Many of these iron alloy reinforcements are submitted to atmospheric corrosion for several hundred years, and the behaviour of these heritage metallic materials has to be predicted for conservation and restoration reasons. In particular, it would be very helpful to establish a degradation diagnosis method. To set up such a diagnosis, it is crucial to consider the chemical and electrochemical reactivities of the microscopic phases present inside the corrosion scales.

Ancient (i.e. several centuries old) corrosion systems are mainly composed of well crystallised goethite α-FeOOH.[1–4] This phase is often embedded with others, such as ferrihydrite ($Fe_5HO_8.9H_2O$), feroxyhyte (δ-FeOOH) and low crystallised maghemite (γ-Fe_2O_3). It has been shown that these latter phases and especially ferrihydrite are electrochemically reactive in the conditions of the atmospheric corrosion.[5] If they are connected to the metal, they play a role in the corrosion mechanisms. Lepidocrocite (γ-FeOOH) and akaganeite (β-FeOOH) can also be locally present, the latter only in the external part of the layer. These reactive phases influence the corrosion mechanisms during the wet–dry cycle of indoor atmospheric environment.[6–8] Phases such as lepidocrocite, maghemite, ferrihydrite, akaganeite, feroxyhyte and even low crystallised goethite can be reduced during the wetting stage instead of atmospheric oxygen,[5,9] causing an important metallic iron corrosion. Other mechanisms strongly linked with the corrosion product phases are also suggested in the literature, such as oxygen migration through the corrosion product layer or local solubilisation of nanometric islets associated with ions migration.[6]

Consequently, it is crucial for corrosion diagnosis to establish the corrosion layer reactivity, linked to the nature and the quantity of constitutive phases.[10,11] Several authors tried to define a reactivity ratio from these parameters (Table 1). The earliest ones only took into account goethite and lepidocrocite. Then, other phases present in the corrosion layer (magnetite, amorphous ferric oxide and akaganeite) were considered in the ratio, but in any case, the intrinsic reduction reactivity of the constitutive phases, especially low crystallised phases,[12] is never taken into account.

[1]Institut de Chimie et des Matériaux Paris-Est, UMR7182, Université Paris-Est, 2 to 8 rue Henri Dunant, Thiais, 94320 Paris, France
[2]Laboratoire Archéomatériaux et Prévision de l'Altération SIS2M UMR3299 CEA/CNRS and IRAMAT UMR5060 CNRS, CEA Saclay, Gif sur Yvette, 91191 Paris, France
[3]Laboratoire Analyse et Modélisation pour la Biologie et l'Environnement, UMR8587, Université d'Evry, rue du Père Jarlan, 91025 Evry, France

*Corresponding author, email monnier@icmpe.cnrs.fr

Published by Maney on behalf of the Institute
Received 12 February 2010; accepted 26 June 2010
DOI 10.1179/147842210X12779093813740

Table 1 Various definitions of 'protectivity ratio' describing stability of corrosion product layer in case of atmospheric corrosion in literature

Reference	Definition of the protectivity ratio
4, 13-15 Miyuki et al. cited by 16	$\frac{\alpha}{\gamma}=\frac{\%_{\alpha-FeOOH}}{\%_{\gamma-FeOOH}}$ †
17	$\frac{\alpha}{\gamma^*}=\frac{\%_{\alpha-FeOOH}}{\%_{\beta-FeOOH}+\%_{\gamma-FeOOH}+\%_{Fe_3O_4}}$
16	$\frac{\alpha+am}{\gamma^*}=\frac{\%_{\alpha-FeOOH}+\%_{amorphous}}{\%_{\beta-FeOOH}+\%_{\gamma-FeOOH}+\%_{Fe_3O_4}}$ ‡
18	$\frac{\alpha^*}{\gamma^{**}}=\frac{\%_{\alpha-FeOOH}+\%_{Fe_3O_4}}{\%_{\beta-FeOOH}+\%_{\gamma-FeOOH}}$

† $\%_{\alpha\text{-FeOOH}}$ = mass fraction of goethite and $\%_{\gamma\text{-FeOOH}}$ = mass fraction of lepidocrocite.
‡ am = amorphous phases detected with XRD.

The present paper brings new results of an electrochemical study of samples taken on a cultural heritage monument to evaluate the reactivity of their rust. These results will be compared in regard to the composition of the rust layer. A new proposal for a reactivity ratio, taking into account the complexity of the layer, will be made and discussed.

Experimental

Samples

The present study is based on samples coming from the iron chains of the Amiens Cathedral (France), submitted to indoor atmospheric corrosion since 1497.[19] In this cathedral, relative humidity and temperature sensors were put at different places to check the homogeneity of the corrosive atmosphere inside the building and evaluate the aspect of the wet–dry cycles. Over a period of 2 years, ranges of variation inside the cathedral are between 50 and 90% for relative humidity and from 10 to 20°C for temperature. The chain is exposed to wet–dry cycles lasting from several days up to several weeks.[6] Four samples were selected for the present study. The compositions of their rust scales, analysed in previous studies, are indicated in Table 2.[12] As several mappings on each sample ensure a statistical view of the determined surface fraction f_S, it can be assimilated to a volume fraction f_V. At four locations, near the sampling zone, rust powder was taken by scratching the artefact until the metallic substrate is seen.

In addition, reference samples of iron oxyhydroxides for the electrochemical measurements have been synthesised following Cornell and Schwertmann's protocols optimised by Lair *et al.*[5] as referenced in Table 3. A difference in crystallinity between the two goethite samples occurs, depending on the synthesis route.

Electrochemical reactivity measurements

The electrochemical reactivity of reference and scratched ancient powders has been tested according to the experimental procedure established by Antony *et al.*[20] The reduction proceeds by imposing a current density of 25 μA mg^{-1} powder, and the potential coulombic charge transient gives information on the reduction behaviour. Galvanostatic reduction experiments are performed in a cell containing Ag/AgCl reference electrode and a platinum wire as a counter electrode (~1 cm^2 surface area). The working electrode is made of graphite mixed with the sample powder (80–20 wt-% respectively to obtain ~10 mg pellets) compressed on a stainless steel grid of 1·25 cm^2 in the surface. Tested samples are not only reference powders but also ancient powders scratched on the Amiens Cathedral chains. The electrolyte used is a deaerated 0·1M NaCl solution, with buffered pH at 7·5 [1,4-piperazinediethanesulfonic acid, disodium salt (Aldrich, St Louis, MO, USA, pH buffering agent with pK_a ~6·8) PIPES solution, 0·05M] and maintained at 25°C.

The iron mass fractions of ferric products X_{Fe} are determined by energy dispersive spectroscopy coupled to an SEM (accelerating voltage of 15 kV; Cambridge Scientific Instruments Ltd, Cambs, UK) and by complexometric titration of Fe(III) with ethylenediaminetetraacetic acid (EDTA), disodic salt (RP Normapur; Prolabo, Paris, France). A weighted quantity of dried solid (mass within the 30–50 mg range) is dissolved in hot 1M HCl (about 5–7 mL). After complete dissolution, 18 MΩ cm nanopure water is added, until a volume of 50 mL is reached. The solution is then thermostated at 45°C, and three drops of sulphosalicylic acid indicator are added. The equivalent volume of EDTA (0·05M) is measured when the colour changes from dark purple to light yellow.

Table 2 Volume fraction of identified phases goethite, ferrihydrite, akaganeite and ferrihydrite on four samples coming from Amiens Cathedral chains: sample names are defined in Refs. 6 and 19

	Goethite f_V	Lepidocrocite f_V	Akaganeite f_V	Ferrihydrite type phase f_V
Am IV E (sample and powder)	0·43±0·15	0·10±0·08	0·11±0·07	0·36±0·13
Am XXVIII c (sample and powder)	0·39±0·04	0·07±0·05	0·19±0·07	0·35±0·13
Am XXX S (sample and powder)	0·28±0·06	0·05±0·02	0·12±0·05	0·55±0·09
Am LX p (sample near the Am LIII powder)	0·35±0·30	0·02±0·02	0·06±0·03	0·57±0·31

Table 3 Description of synthesised iron oxyhydroxides

Reference phase	Origin
Goethite G1	Well crystallised phase synthesised by alkaline reprecipitation of ferrihydrite at 70°C during 3 days
Goethite G2	Low crystallised goethite synthesised by oxidation of $FeCl_2.4H_2O$ in a $NaHCO_3$ solution
Lepidocrocite	Synthesis by oxidation of a $FeCl_2.4H_2O$ solution at pH of ~6·8
Two-line ferrihydrite	Synthesis by precipitation with NaOH of a $Fe(NO_3)_3.9H_2O$ solution until pH of ~8
Akaganeite	Synthesis by aging of a $FeCl_3.6H_2O$ solution at 40°C during 8 days

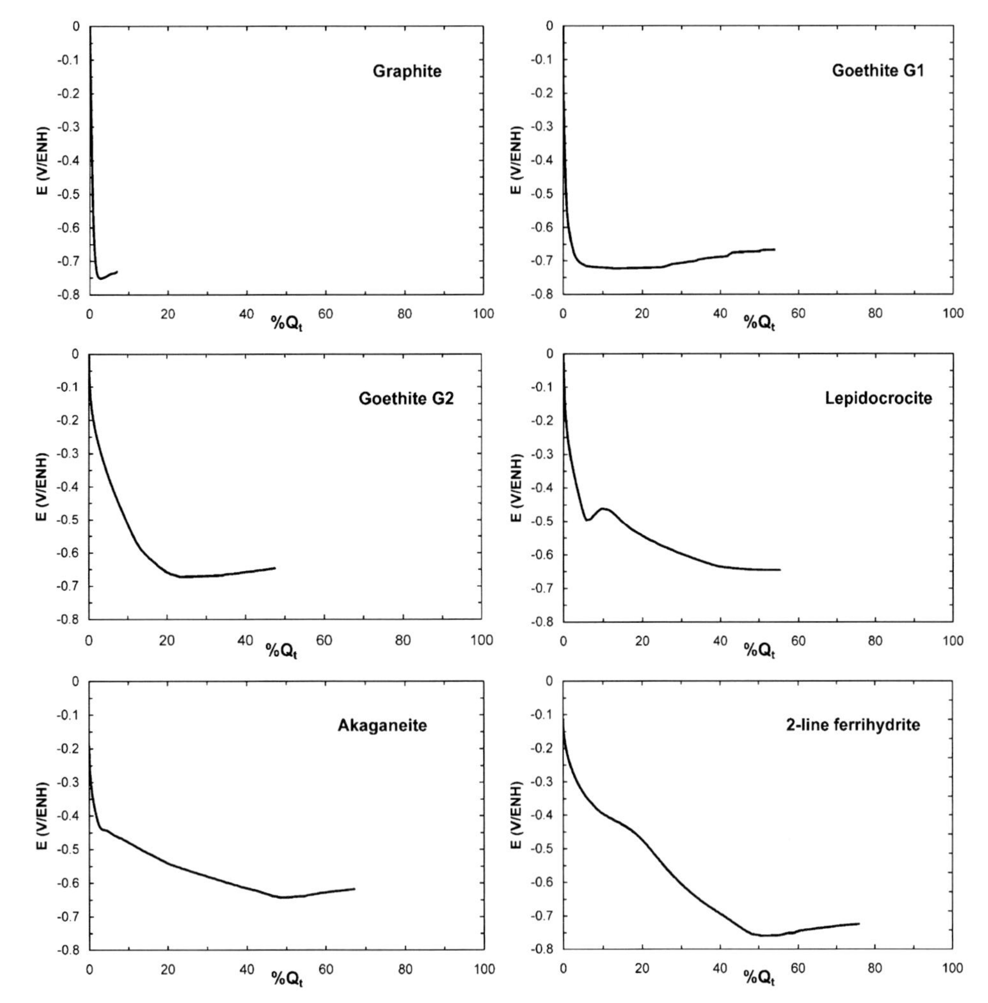

1 Electrochemical reduction curves giving potential E normalised to normal hydrogen electrode NHE versus reduction ratio $\%Q_t$ for graphite and for reference powders: G1, well crystallised goethite; G2, low crystallised goethite; Lep, lepidocrocite; Aka, akaganeite and Fh, ferrihydrite

Results and discussion

Electrochemical reactivity

The reduction curves obtained on reference powders (Fig. 1) show the potential E versus the reduction ratio $\%Q_t$ described by Lair *et al.*[5] as follows

$$\%Q_t = \frac{i_c t M}{m \Im X_{Fe}} \times 100\%$$

where i_c is the cathodic current imposed for the reduction (A), t is the time (s), M is the molar mass of iron (55·85 g mol^{-1}), $\Im$ is the Faraday constant, m is the sample mass in the composite working electrode and X_{Fe} is the Fe(III) mass fraction in the samples, determined by EDTA analysis or by energy dispersive spectroscopy coupled to an SEM and given in Table 4.

Table 4 Iron mass fraction X_{Fe} in both reference and ancient powder samples*, %

Compound	X_{Fe}, %
Goethite 1	63
Goethite 2	55
Lepidocrocite	56
Akaganeite	52
Ferrihydrite	51
Am IV powder	52
Am XXVIII powder	52
Am XXX powder	46
Am LIII powder	38

*The experimental errors on X_{Fe} are under 3%.

Two parameters (Table 5) are first extracted from the reduction curves in order to compare the reactivities of the phases:

(i) $\%Q_\tau$ is the $\%Q_t$ value at the lowest potential on the reduction curve. It is linked to the total consumed charge involved in the reduction in the compound

(ii) $E_{\tau/2}$ is the half reduction potential corresponding to the potential measured at half the $\%Q_\tau$ value.

More details about this methodology can be found in the literature.[5]

The curve of pure graphite electrode is a reference of a non-reactive electrode: after the sharp falling down of the potential, the increase at the end of the process corresponds to the reduction in the electrolyte. The curve of goethite G1, the more crystallised phase, is close to that of graphite with a fast potential drop showing a low reactivity ($\%Q_\tau$=8%). Moreover, goethite G1 half reduction potential value [$E_{\tau/2}$=−0·70 V(NHE)] is cathodic compared to that of metallic iron [E=−0·62 V(NHE)], so a coupling reaction between well crystallised goethite and metallic iron would be quite difficult. The G2 powder seems to be slightly more reactive, with a curve showing a higher value of $\%Q_\tau$=22% and a half potential value more anodic [$E_{\tau/2}$=−0·54 V(NHE)]. This difference could be related to the difference in crystallinity. This result indicates that although goethite is often considered as a non-reactive

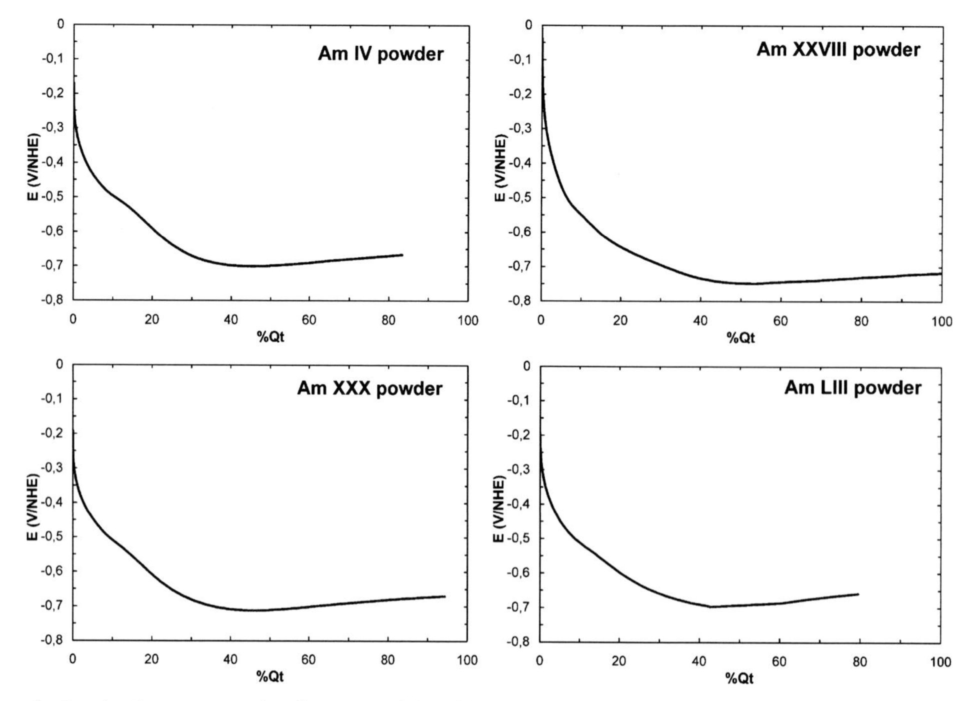

2 Electrochemical reduction curves showing potential E (NHE) *versus* reduction ratio $\%Q_t$ of four ancient samples

phase in the literature, it can be reduced during the wet stage of the relative humidity cycle, giving iron corrosion.

For atmospheric corrosion, whereas the Cl seems to precipitate locally at the outer part of the layer,[1] the akaganeite phase could act in alteration mechanisms as it presents quite a high reduction reactivity, with a $\%Q_\tau$ of 46% and a half reduction time potential of −0·55 V(NHE). It is shown here that lepidocrocite is a reduction reactive phase with a $\%Q_\tau$ of 42% and an $E_{t/2}$ of −0·55 V(NHE). It can therefore act in the alteration process. This result is in contradiction with a recent study combining electrochemistry and Raman microspectroscopy by Bernard and Joiret.[21] In fact, these authors used high reduction current values, which provoked the reduction in both lepidocrocite and water. In the present conditions, with a better choice for current values, the reduction in lepidocrocite unambiguously occurs at potentials less cathodic than water. Ferrihydrite is the most reducible phase in the system, with a $\%Q_\tau$ value of 50%. In summary, phases such as lepidocrocite, akaganeite and ferrihydrite have to be considered as reactive ones, in good agreement with former studies,[5] whereas some phases, such as well crystallised goethite, are stable ones.

In order to discuss the reactivity of the tested powders, it appears that both the reduction factor $\%Q_\tau$ and the half reduction potential $E_{\tau/2}$ should be considered. An electrochemical reactivity factor ρ_i is defined as follows

$$\rho_i = \frac{Q_{\tau,i}\Delta E_i}{9{\cdot}40}$$

where ΔE_i is the difference for a given powder between $E_{\tau/2}$ and E_{water}. This latter parameter is the minimum potential of each curve, i.e. the potential where water reduction begins. The values of E_{water} and ΔE_i are given in Table 5 for each reference phase. For the most reactive phase, ferrihydrite, the $Q_{\tau,i}\Delta E_i$ value of 9·40 has to be compared to the value of 0·16 obtained for the most stable phase, goethite G1. To express the electrochemical reactivity factor, a normalisation factor of 9·40 was used. Thus, the ρ factor presents values between 0·02 and 1.

Nevertheless, it must be kept in mind that several factors can influence the phase reduction reactivity, such as crystallinity degree, solubility and the presence of minor elements. These parameters can vary a lot depending on the ancient samples. Thus, it is worthwhile to test directly the reactivity of samples constituted of ancient rusts. The reduction curves of ancient samples

Table 5 Reduction ratio Q_τ, half reduction potential $E_{\tau/2}$, water reduction potential E_{water}, potential difference ΔE, $Q_\tau E_{\tau/2}$ factor and ρ values* for reference powders and historical samples

Compound	Q_τ, %	$E_{\tau/2}$, V(NHE)	E_{water}, V(NHE)	ΔE, V(NHE)	$Q_{\tau,i}\Delta E_i$	ρ_i
Goethite 1	8	−0·70	−0·72	0·02	0·16	0·02
Goethite 2	22	−0·54	−0·67	0·13	2·86	0·30
Lepidocrocite	42	−0·55	−0·65	0·10	3·99	0·42
Akaganeite	46	−0·55	−0·64	0·09	4·09	0·44
Ferrihydrite	50	−0·47	−0·66	0·19	9·40	1·00
Am IV powder	40	−0·59	−0·70	0·11	4·32	0·46
Am XXVIII powder	48	−0·67	−0·75	0·08	3·94	0·42
Am XXX powder	41	−0·61	−0·71	0·10	4·02	0·43
Am LIII powder	43	−0·61	−0·70	0·09	3·96	0·42

*The experimental errors on Q_τ are under 4%, and the ones on potential measurements are under 5% that means errors on the electrochemical reactivity factor ρ_i values under 14%.

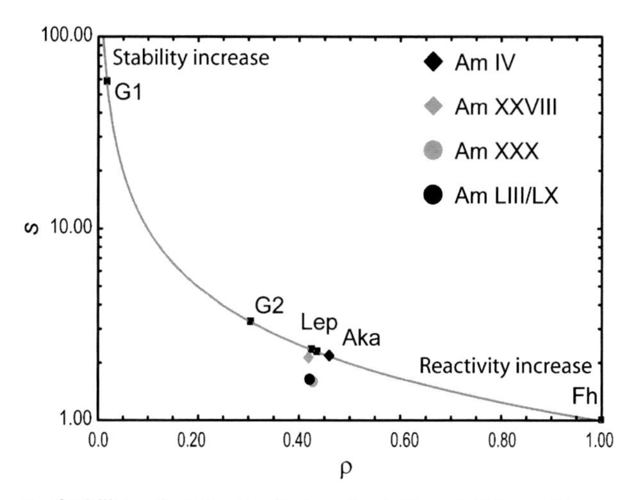

3 Stability abacus worked out from stability ratio s values determined on massive samples and electrochemical reactivity factor ρ values determined on reference and ancient powders: squares stand for samples entirely composed of reference phases, and other points stand for Amiens Cathedral site samples; grey line shows envelop $s=1/\rho$ curve

are shown in Fig. 2, and their Q_τ, $E_{\tau/2}$ and ρ values are reported in Table 5. The curves present more or less pronounced reduction plateau but show a similar global behaviour with a low decrease in potential and a calculated alpha factor around 0·43. This value is close to the one of pure lepidocrocite. Thus, concerning their electrochemical behaviour, the corrosion products of Amiens Cathedral can be considered as equivalent and relatively reactive.

New stability index for reliable degradation diagnosis

From the previous paragraph, it appears that reactivity is different from one phase to another. That point was never taken into account in previous studies. Thus, a new 'stability ratio' named s that considers different reactivities of the reference phases is proposed. Weighting coefficients for each reference phase are introduced in the ratio definition. In the first approach, these coefficients could be based on the electrochemical reactivity factor ρ_i and the volume fraction for each constitutive phase so the new ratio s is defined as

$$s=\frac{1}{\sum_i \rho_i f_{V,i}}$$

where i stands for each identified phase (goethite, akaganeite, lepidocrocite and ferrihydrite), $f_{V,i}$ represents the volume fraction of the ith phase and ρ_i is the electrochemical reactivity factor of the ith reference phase.

The main difficulty consists in the evaluation of the phase proportion in the corrosion layer and especially the less crystallised ones. For that, the semiquantification program LADIR-CAT[12] (Table 2) can be used. However, this program does not allow distinguishing ferrihydrite from other low crystallised phases. In the first approach, all reactive phases are considered to be ferrihydrite. Consequently, as this is the more reactive phase, the protective index could be slightly underrated. Moreover, the LADIR-CAT program is currently not able to distinguish between different kinds of goethite. Previous studies showed that goethite from the site of Amiens Cathedral is globally well crystallised;[1] thus, only the well crystallised goethite in the s factor determination will be considered here.

Table 6 shows the s factor values for reference and ancient samples. Two groups of data, Am IV and Am XXVIII on the one hand (s of ~2·15) and Am XXX and Am LX on the other hand (s of ~1·62), can therefore be distinguished. This difference may be related to the higher reactive ferrihydrite content in the two latter samples (Table 2).

For a hypothetical low reactive corrosion product layer entirely made of goethite G1, the s value would be 58·75. Conversely, a very reactive hypothetical layer entirely made of ferrihydrite would have an s value of 1·00. Thus, as for the ρ factor, the corrosion layers of Amiens Cathedral can be considered as relatively reactive.

By combining the stability ratio s values determined from quantitative phase measurements on massive samples and the electrochemical reactivity factor ρ measured from electrochemical study of the ancient powders, a stability abacus for indoor atmospheric corrosion systems can be drawn (Fig. 3). Once again, it can be observed that the reactivity values are in the same order of magnitude as that of pure reactive phases.

If the different ways to express the reactivity of corrosion product layer are considered, the authors propose here to mainly take into account the reduction reactivity. The values of ρ obtained on the four Amiens Cathedral powders are very close, between 0·42 and 0·46, and lead to a classification of the Amiens Cathedral site samples as reactive ones. This reactivity is also stated from s stability ratio values and mainly results from the large ferrihydrite content of these samples; it can be noted that, as explained before, the LADIR-CAT program actually considers all the low crystallised phases as ferrihydrite. These results are consistent with previous studies, which stated a decrease in reduction reactivity with time and classified Amiens samples as reactive.[20] To better define intermediary steps in the stability scale, it would be necessary to study samples coming from other sites and mixtures of reference phases with very different compositions, from pure well crystallised goethite to extremely reactive phases like ferrihydrite or probably even more reactive phases. However, the determined s factor only takes into account the electrochemical reactivity of the different phases, although several corrosion mechanisms highlight the importance of the connectivity and the conductivity of the different phases. A final stability factor should consequently include a second weighting

Table 6 s stability ratio for ancient samples

	s stability factor
Goethite 1	58·75
Goethite 2	3·29
Lepidocrocite	2·36
Akaganeite	2·30
Ferrihydrite	1·00
Am IV sample	2·18
Am XXVIII sample	2·13
Am XXX sample	1·59
Am LX sample (near LIII powder)	1·64

coefficient illustrating the connectivity/conductivity of the different phases with the metal.

This abacus enables to distinguish samples that have the same global electrochemical behaviour but different compositions. In particular, the two groups of samples, Am XXX and Am LX on the one hand and Am IV and Am XXVIII on the other hand, show similar ρ values (powders) but lower s values (massive samples). This is due to their difference in composition. This example illustrates the interest of achieving the two types of experiment, electrochemistry on the powder to get the ρ value and quantification on the massive sample to get the s one.

Finally, the abacus can give a degradation diagnostic for a rust layer under different sampling constraints from its two inputs, s or ρ. If sampling of several samples of corrosion product layers, including the metallic core, is possible, the protectivity ratio s could be based on micro-Raman analysis of these samples, to avoid focalising on a particular case. For the same reasons, several Raman mapping should be undertaken on each specimen. If powder samples can be scrapped from the artefact, the ρ value can be determined from electrochemical study. Moreover, if the massive sampling is impossible, a global composition can also be assessed from the powder sample and used for the determination of s value, even if a risk of over-representation of the external layers may exist in this case.

Conclusion

In the present paper, several samples from Amiens Cathedral (the fifteenth century) chains are investigated. The authors propose two indexes for the evaluation of iron rust layer reactivity/stability. The first one, the electrochemical reactivity ratio ρ, is worked out from the galvanostatic reduction transient of the powder and takes into account both the reduction factor and the potential decrease. The second one s is determined from mass fraction on massive samples and ρ values of phases constituting the powder. The ρ and s factor association in an abacus enables to propose a degradation diagnosis for long term indoor atmospheric corrosion of iron, taking into account both powder and massive sampling.

To improve this diagnosis, two ways have to be explored: the first is studying other powders with very different compositions to observe the evolution of α and s factors and second is taking into account the connectivity and the conductivity of the different phases.

Acknowledgements

This study was supported by the PNRC Project 'Mise en place d'une méthode de diagnosis de la corrosion atmosphérique' of the Ministère de la Culture et de la Communication, by the GdR 3174 of CNRS 'ChimARC' and the ARCOR Project supported by the French National Research Agency. The authors friendly thank C. Paris and L. Bellot-Gurlet (LADIR) for their precious help in the micro-Raman mapping acquisitions.

References

1. J. Monnier, D. Neff, S. Réguer, P. Dillmann, L. Bellot-Gurlet, E. Leroy, E. Foy, L. Legrand and I. Guillot: 'A corrosion study on the ferrous medieval reinforcement of the Amiens Cathedral. Phase characterisation and localisation by various microprobes', *Corros. Sci.*, 2010, **52**, 695–710.
2. R. Balasubramaniam: 'On the corrosion resistance of the Dehli iron pillar', *Corros. Sci.*, 2000, **42**, 2103–2129.
3. T. Kamimura and N. Stratmann: 'Mössbauer spectroscopic study of rust formed on a weathering steel exposed for 15 years in an industrial environment', *Mater. Trans.*, 2000, **41**, (9), 1208–1215.
4. M. Yamashita, H. Miyuki, Y. Matsuda, H. Nagano and T. Misawa: 'The long term growth of the protective rust layer formed on weathering steel by atmospheric corrosion during a quarter of a century', *Corros. Sci.*, 1994, **36**, (2), 283–299.
5. V. Lair, H. Antony, L. Legrand and A. Chausse: 'Electrochemical reduction of ferric corrosion products and evaluation of galvanic coupling with iron', *Corros. Sci.*, 2006, **48**, 2050–2063.
6. J. Monnier: 'Corrosion atmosphérique sous abri d'alliages ferreux historiques. Caractérisation du système, mécanismes et apport à la modélisation', PhD thesis, Université Paris-Est, Paris, France, 2008, available at: http://tel.archives-ouvertes.fr/tel-00369510/fr/
7. M. Stratmann and J. Müller: 'The mechanism of the oxygen reduction on rust-covered metal substrates', *Corros. Sci.*, 1994, **36**, (2), 327–359.
8. U. R. Evans and C. A. J. Taylor: 'Mechanism of atmospheric rusting', *Corros. Sci.*, 1972, **12**, (3), 227–246.
9. M. Stratmann and K. Hoffmann: '*In situ* Mössbauer spectroscopic study of reactions within rust layers', *Corros. Sci.*, 1989, **29**, (11–12), 1329–1352.
10. J. T. Keiser, C. W. Brown and R. H. Heidersbach: 'Characterization of the passive film formed on weathering steels', *Corros. Sci.*, 1983, **23**, (3), 251–259.
11. T. Okada, Y. Ishii, T. Mizoguchi, I. Tamura, Y. Kobayashi, Y. Takagi, S. Suzuki, H. Kihira, M. Itou, A. Usami, K. Tanabe and K. Masuda: 'Mössbauer studies on particle volume distribution of α-FeOOH in rust formed on weathering steel', *Jpn J. Appl. Phys.*, 2000, **39**, (6A), 3382–3391.
12. J. Monnier, D. Baron, L. Bellot-Gurlet, D. Neff, I. Guillot and P. Dillmann: 'A methodology for Raman structural quantification imaging and its application to iron indoor atmospheric corrosion products', *J. Raman Spectrosc.*, 2010, DOI: 10.1002/jrs.2765.
13. H. Kishikawa, H. Miyuki, S. Hara, M. Kamiya and M. Kamashita: Proc. 45th Japan Conf. on 'Materials and environments', 115; 1998, Tokyo, Japan Society of Corrosion Engineering.
14. H. Miyuki, M. Yamashita, M. Fujiwara and T. Misawa: 'Titre japonais', *Zairyo-to-Kankyo*, 1998, **47**, 186–192.
15. K. Kashima, S. Hara, H. Kishikawa and H. Miyuki: 'Evaluation of protective ability of rust layers on weathering steels by potential measurement', *Corros. Eng.*, 2000, **49**, 25–37.
16. H. Kihira, T. Misawa, T. Kusunoki, K. Tanabe and T. Saito: 'How to use the composition ratio index obtained by internal standard quantitative X-ray diffraction analysis to evaluate the state of rust on weathering steel', *Corros. Eng.*, 1999, **48**, 979–987.
17. P. Dillmann, F. Mazaudier and S. Hoerle: 'Advances in understanding atmospheric corrosion of iron. I. Rust characterisation of ancient ferrous artefacts exposed to indoor atmospheric corrosion', *Corros. Sci.*, 2004, **46**, (6), 1401–1429.
18. S. Hoerlé, F. Mazaudier, P. Dillmann and G. Santarini: 'Advances in understanding atmospheric corrosion of iron. II. Mechanistic modelling of wet–dry cycles', *Corros. Sci.*, 2004, **46**, (6), 1431–1465.
19. E. Lefèbvre: 'La place et le rôle des métaux dans l'architecture gothique à travers l'exemple de la cathédrale d'Amiens. Étude de cas: le chaînage du triforium de la cathédrale d'Amiens', in 'Master d'Archéologie médiévale', 127; 2006, Amiens, Université d'Amiens.
20. H. Antony, S. Perrin, P. Dillmann, L. Legrand and A. Chaussé: 'Electrochemical study of indoor atmospheric corrosion layers formed on ancient iron artefacts', *Electrochim. Acta*, 2007, **52**, (27), 7754–7759.
21. M. C. Bernard and S. Joiret: 'Understanding corrosion of ancient metals for the conservation of cultural heritage', *Electrochim. Acta*, 2009, **54**, 5199–5205.

Characterisation of corrosion layers formed on ferrous archaeological artefacts buried in anoxic media

M. Saheb*[1,2], D. Neff[1], J. Demory[1], E. Foy[1] and P. Dillmann[1,3]

In the context of the *in situ* conservation and preservation of archaeological artefacts, the long term corrosion mechanisms of iron in anoxic soils are studied. To this purpose, a first step is the characterisation of the corrosion layers formed on archaeological artefacts provided from the archaeological site of Glinet (the sixteenth, Normandy, France). On all the corrosion systems formed on artefacts, the main phases constitutive of the corrosion layer are siderite ($FeCO_3$), an iron carbonate containing hydroxide groups [probably chukanovite $Fe_2(OH)_2CO_3$] and magnetite (Fe_3O_4). Furthermore, the arrangement of these phases reveals three corrosion distribution types with corresponding corrosion pattern diagrams.

Keywords: Anoxic corrosion, Archaeological artefacts, Corrosion layer characterisation, Microbeam techniques

This paper is part of a special issue on corrosion of archaeological and heritage artefacts

Introduction

Understanding the corrosion processes occurring on archaeological iron buried in anoxic environments will offer guidance for predicting its continuing survival in these conditions. This can be useful when object numbers and resource issues dictate the need for *in situ* preservation of archaeological finds.[1] In the past, this has included iron in anoxic environments.[2–5] In these conditions iron corrosion is expected to be slow, but there is limited knowledge of corrosion mechanisms and rates. Examination and categorisation of corrosion profiles on iron from anoxic environments will contribute to elucidate corrosion processes and this will allow better interpretation of thermodynamic data. In future it may be possible to predict the likely corrosion state of archaeological iron from assessment of the burial environment. As a first step towards this goal characterisation of the iron corrosion profiles must be linked to the burial environment, thereby creating an understanding of the burial corrosion system.

Previous studies on archaeological sites presenting an anoxic burial environment (Nydam Mose, Denmark;[6–8] Fiskerston, UK;[4,5] Glinet, France[8–10] and Saint-Louis, France[8]) reveal that siderite $FeCO_3$ is the main phase identified on the corrosion layers formed on ferrous artefacts. In addition to siderite, depending on the environmental conditions, other phases have also been found, as other carbonates such as $Fe_2(OH)_2CO_3$, named chukanovite[11–13] (Saint-Louis). Oxide phases have also been identified, as magnetite (Fe_3O_4). The sulphur containing phases greigite Fe_3S_4, pyrite (FeS_2) and mackinawite ($Fe_{1+x}S$) were also reported at the archaeological site of Fiskerston.[7] This latter phase, sometimes partially oxidised has also been reported at Glinet.[14] In the presence of phosphorus, vivianite $[Fe_3(PO_4)_2.8H_2O]$ has been sometimes identified in the corrosion products.[7] On several ferrous archaeological artefacts, Neff[10,15] developed a terminology to describe the corrosion products patterns and distribution on ferrous archaeological objects, which will be used here. The corrosion product pattern is divided into four parts. The metallic substrate is made of iron that often contains slag inclusions.[16] The dense product layer (DPL) contains the corrosion phases. The transformed medium (TM) contains markers from the soil (i.e. quartz and calcite grains) but also high iron amounts due to iron migration from the corrosion products. This iron amount decreases progressively from the DPL to the soil. Lastly, surrounding the TM, the soil is characterised by an elemental composition which has not been influenced by iron corrosion.

The aim of this paper was to characterise the corrosion layers formed on artefacts from an anoxic zone within the archaeological site of Glinet and to propose formation mechanisms for the identified phases. Parameters of the burial soil are published elsewhere showing anoxic conditions. Furthermore, the major ions present in the porewater are calcium and carbonate, corresponding to a calco-carbonated electrolyte.[17] On the archaeological artefacts complementary microbeam techniques have been used to investigate the corrosion system on transverse sections allowing identification of phases in the corrosion layer.

[1]LAPA/SIS2M, UMR 3299 CEA/CNRS, CEA Saclay 91191 Gif-sur Yvette Cedex, France
[2]ANDRA, Parc de la Croix Blanche, 1–7 avenue Jean Monnet, 92298 Chatenay Malabry cedex, France
[3]Institut de Recherches sur les Archéomatériaux, UMR 5060 CNRS, France

*Corresponding author, email mandana.saheb@cea.fr

Published by Maney on behalf of the Institute
Received 2 January 2010; accepted 19 June 2010
DOI 10.1179/147842210X12772898886889

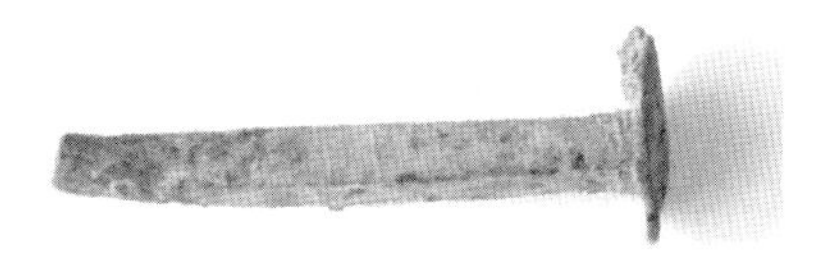

1 Photograph of nail GL07-35 provided from archaeological site of Glinet

Materials and methods

Site and samples

The analysed iron artefacts originate from the site at Glinet (Seine Maritime, France), where iron ore was reduced during the sixteenth century.[18] Sixteen nails have been extracted from a single waterlogged zone. Environmental parameter measurements confirm that this zone is anoxic.[17] Nails were excavated between July 2007 and July 2008 from a depth of ~1 m (see Table 1 and Fig. 1). Whenever possible, they were excavated with their surrounding soil still attached to avoid any disturbance of their corrosion layers. Storage in absolute ethanol prevented access of oxygen and further corrosion by removing water.

Sample preparation

The sampling was performed in a glove box continually purged with nitrogen to avoid any contact with air, as this may cause fresh corrosion. The items were embedded in epoxy resin and cross-sectioned in a nitrogen atmosphere using a diamond wire saw and oil as a lubricant. The sections were polished under ethanol with SiC papers and diamond paste to grain size of 3 μm.

Characterisation methods

All the samples were observed by optical microscope, and scanning electron microscopy (SEM) coupled with X-ray microanalysis energy dispersive spectroscopy (EDS) with an electron acceleration voltage of 15 kV. Raman microspectroscopy was used to identify and locate precisely the phases inside the corrosion layer (Invia Reflex Renishaw spectrometer, 532 nm; spot size (×50 objective): 3 μm; spectral resolution: 2 cm^{-1}; power filtered down to 100 μW, acquisition time: 300 s). Phase identification was performed using a comparison to reference spectra.[15,19,20] A spectral window between 200 and 1600 cm^{-1} was observed, except for hydroxycarbonates, for which a window between 200 and 3600 cm^{-1} has been selected in order to reach the hydroxide vibration bands at around 3300–3600 cm^{-1}.

X-ray microdiffraction (μXRD) was also used to complete the structural characterisation. This method was performed on a rotating anode generator (monochromatic beam of 17·48 keV focused to a 30×30 μm surface by Xenocs FOX 2D MO 25_25P diffraction optic). Data process was realised thanks to FIT2D[21] and EVA softwares associated to the JCPDS database.

Table 1 Archaeological samples provided from site of Glinet

Samples excavated in 2007	Corrosion pattern type	Samples excavated in 2008	Corrosion pattern type
GL07-29	2	GL08-13	1
GL07-30	2	GL08-14	2
GL07-32	2	GL08-15	1
GL07-35	1	GL08-16	1
GL07-43	2	GL08-17	3
GL07-44	2	GL08-18	2
GL07-45	1	GL08-20	1
GL07-46	2		
GL07-50	1		

Results

As the archaeological artefacts were buried in complex anoxic environments containing local environmental variations, their corrosion layers are heterogeneous. Three corrosion product patterns have been established depending on the phase location inside the layers (Table 1). For all artefacts, corrosion layer thickness is between 30 and 500 μm, with a mean of 120 μm (see Figs. 2 and 3). Cracks and pores have been observed parallel and perpendicular to the metallic interface through the entire layer. The DPL contains iron (up to 60 wt-%) and oxygen (up to 40 wt-%) corresponding to iron corroded phases. Moreover, low amounts of calcium, phosphorus and sulphur (below 5 wt-%) have been detected by EDS in the DPL (see Fig. 4). Their presence is due to the migration of soil elements as none of these elements has been identified in the metal using EDS. Whatever the corrosion pattern, some zones appearing light grey under optical microscope have been identified on the outer part of the DPL (see Fig. 5) corresponding to sulphur containing phases such as mackinawite[22] identified by Raman spectroscopy. More details about their characterisation and formation are published elsewhere.[14]

Type I corrosion pattern (Fig. *6a*)

Figure 2 presents a microphotograph of the corrosion system developed on the sample GL08-16 with the associated Raman spectra. The main phase in the DPL is siderite (dark zones). The light layer located on the outer part of the DPL (thickness between 5 and 30 μm) has been identified as magnetite. The TM is composed of a calcite layer in contact with the magnetite layer and of an outer zone that contains markers from the soil. Among them, silicon, phosphorus, aluminium, calcium and sulphur are present (<6 wt-%). The concentration of these elements in the TM is heterogeneous and can sometimes be over the one measured in the DPL. Moreover a high iron amount (up to 40 wt-%) has been detected.

Type II corrosion pattern (Fig. *6b*)

Figure 3 shows a microphotograph of the corrosion system developed on the sample GL08-18 with the associated microdiffractograms. As in type I corrosion pattern, the dark matrix in the DPL is mainly composed of siderite. Additionally, some zones located near the metallic interface contain another iron carbonate. This phase has been studied using a coupling of Raman microspectroscopy (see Fig. 5) and μXRD (see Fig. 3). Microdiffractograms obtained at this location are in good agreement with the JCPDS file of an iron hydroxycarbonate $Fe_2(OH)_2CO_3$ also called chukanovite,[11,12]

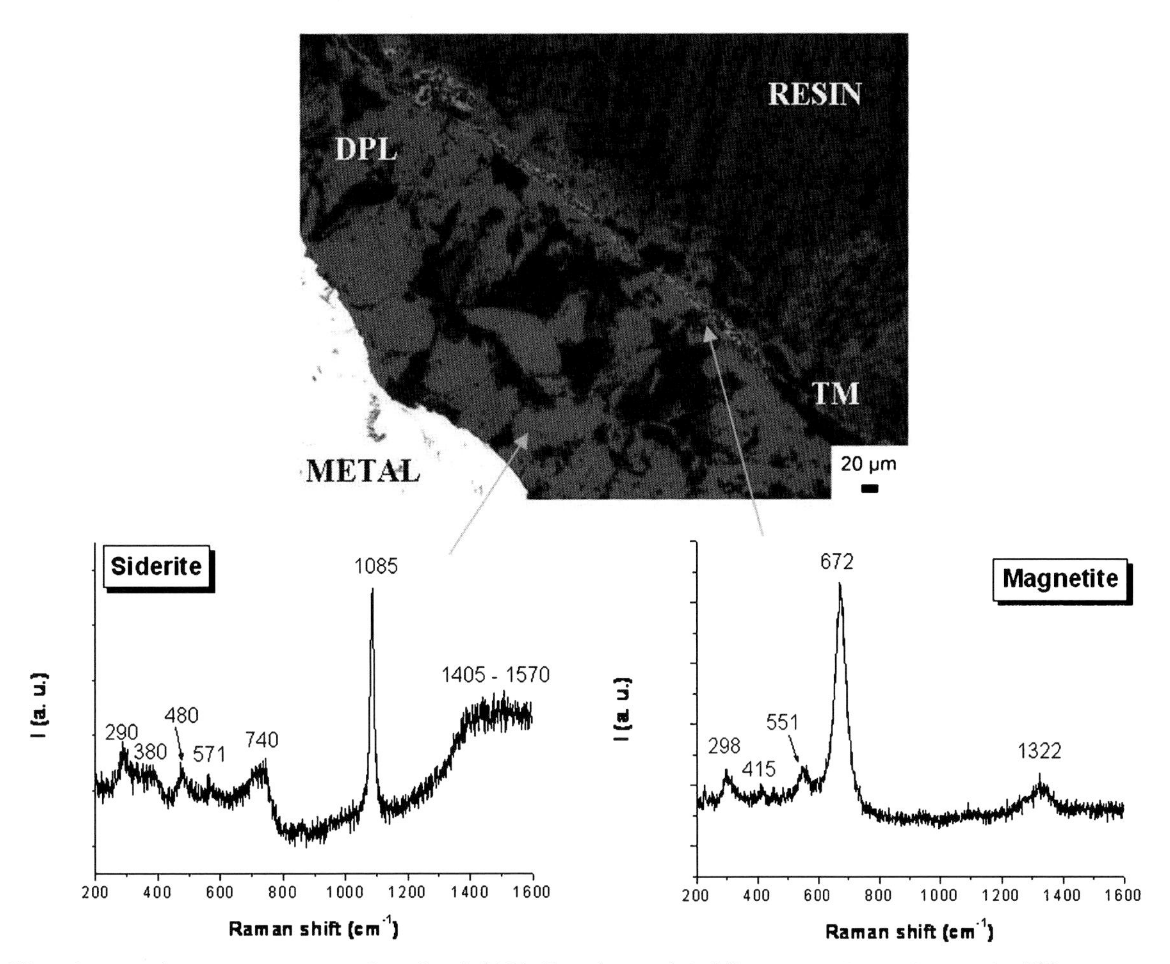

2 Microphotograph on transverse section of nail GL08-16 and associated Raman spectra on two spots, 532 nm

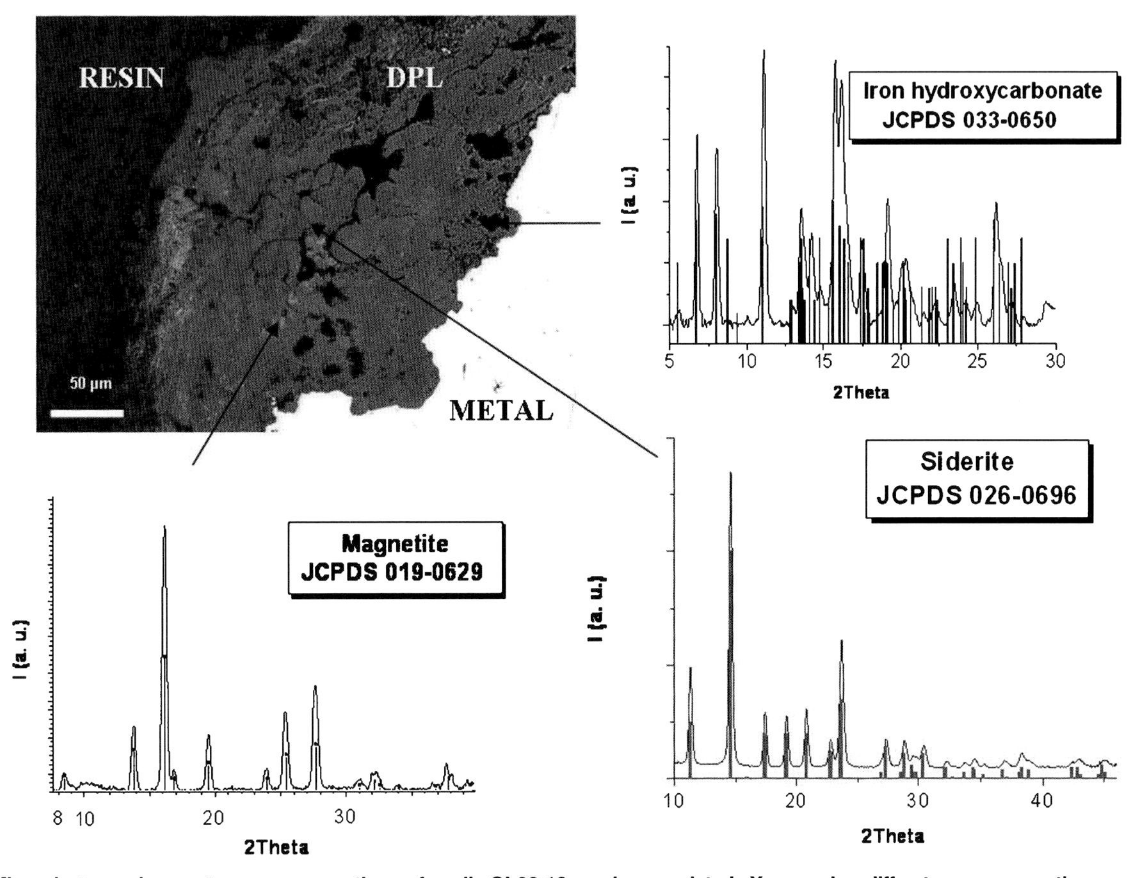

3 Microphotograph on transverse section of nail GL08-18 and associated X-ray microdiffractograms on three zones (20 × 20 μm), 17 keV

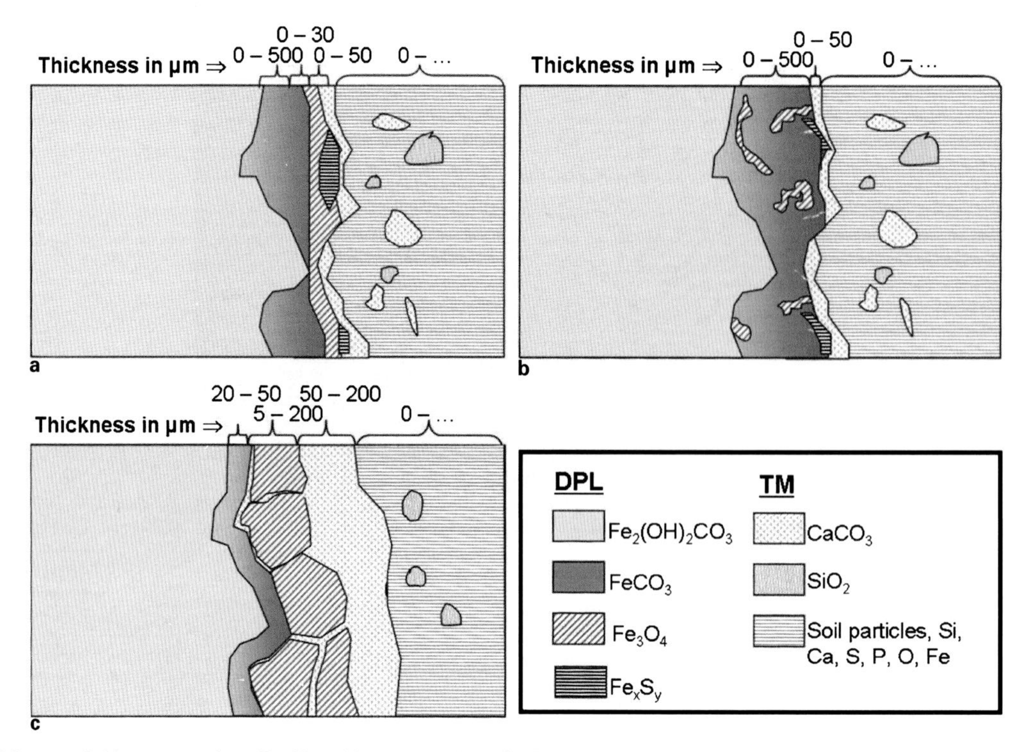

a SEM image; *b* X-ray mapping, Fe K_α; *c* X-ray mapping, O K_α; *d* X-ray mapping, Ca K_α; *e* X-ray mapping, Si K_α

4 Image (SEM) on nail GL08-17, backscattering mode and associated EDS mapping

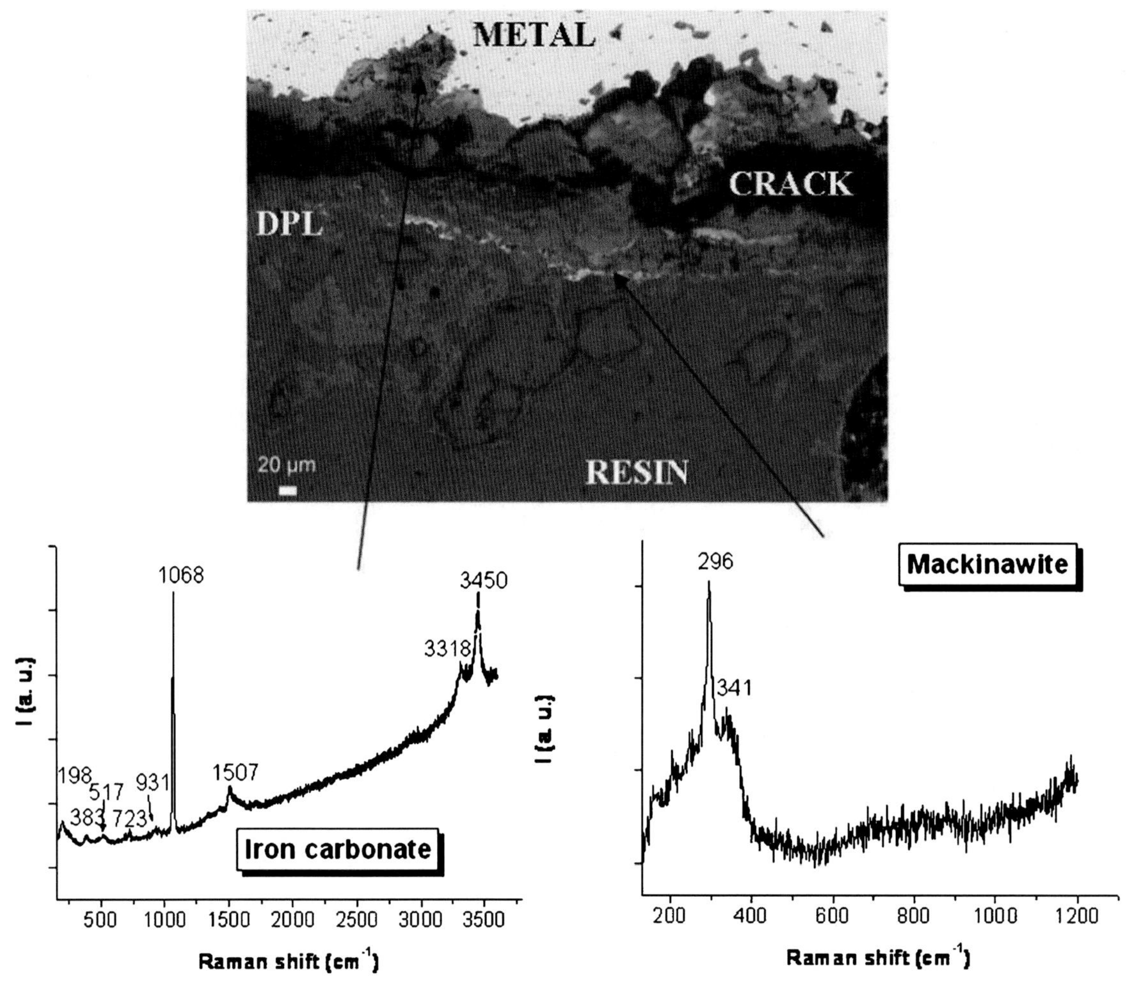

5 Microphotograph on transverse section of nail GL07-43 and associated Raman spectrum on two spots, 532 nm

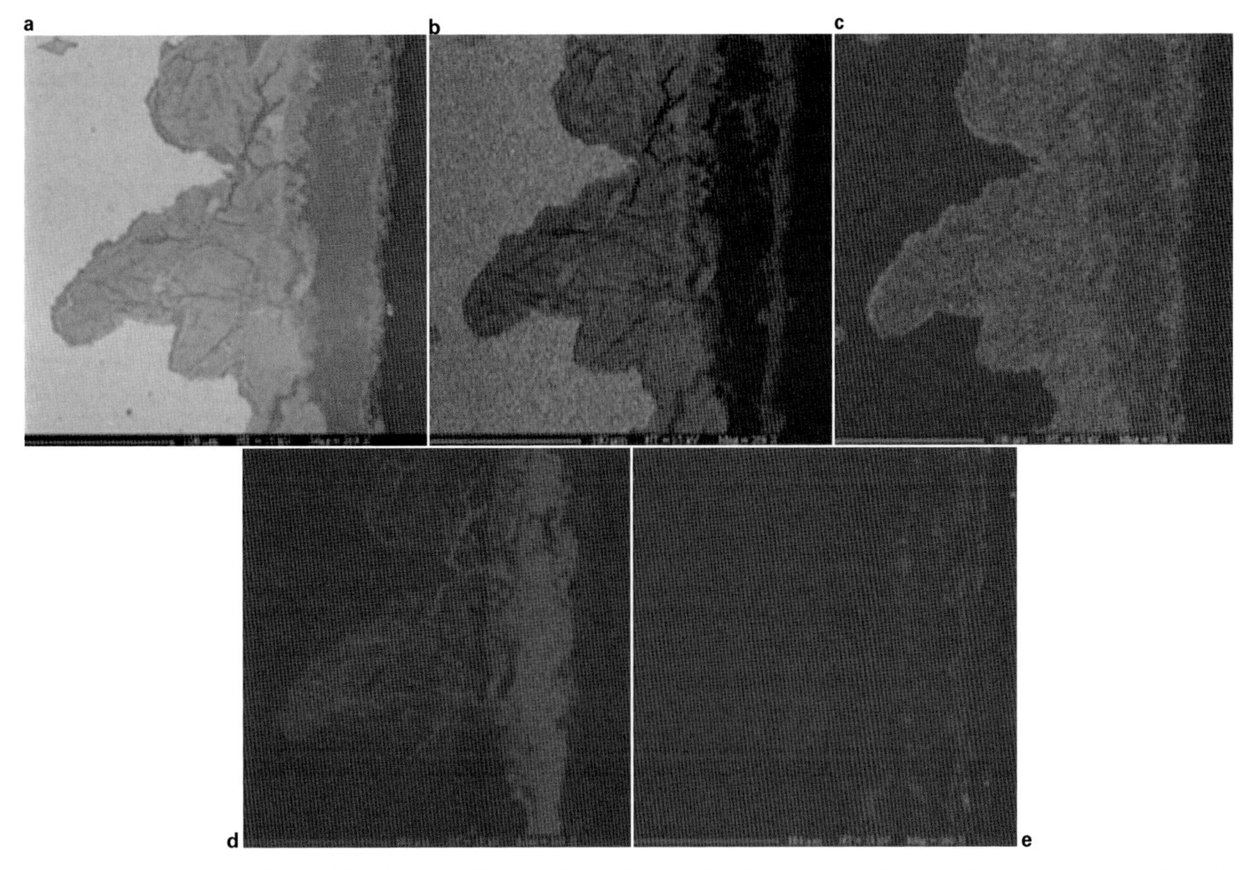

6 Diagram of *a* type I, *b* type II and *c* type III corrosion patterns identified on artefacts provided from archaeological site of Glinet

despite the fact that some diffraction line intensities are not respected. This could be explained by an atomic substitution of iron by calcium in the chukanovite structure, but this requires further investigation in order to confirm this hypothesis. In the Raman spectra, carbonate and hydroxide band groups have been identified at 1065–1070 and 3310–3460 cm^{-1} representing hydrated or hydroxy-carbonate, which is in good agreement with the hypothesis of the occurrence of $Fe_2(OH)_2CO_3$.[8] Inside this dark matrix, lighter zones (between 1 and 10 μm) are present. These have been identified as magnetite by μXRD. The TM presents the same profile as the one identified on the type I corrosion pattern.

Type III corrosion pattern (Fig. 6*c*)

Figure 4 shows a SEM microphotograph in back-scattering mode of the corrosion system developed on the sample GL08-17 with the associated EDS analyses. It reveals that the DPL is divided into two zones. The internal part lies between 10 and 50 μm and contains iron, oxygen and calcium. Siderite has been identified in this zone using Raman spectroscopy. The external part identified as magnetite, presents a variable thickness between 20 and 500 μm. It contains cracks parallel and perpendicular to the interface. These cracks contain calcite, as it has been revealed by coupling Raman microspectroscopy and SEM-EDS, showing respectively a carbonate phase with a main peak at 1080 cm^{-1} and a high calcium amount (20 wt-%). The presence of calcite may be due to the calco-carbonated porewater.

Discussion

In the DPL, three main phases have been identified. Two of them are carbonates: siderite and another ferrous carbonate, probably chukanovite. The third one is magnetite. Furthermore, sulphur containing phases have also been identified on the external part of the DPL.[14]

Magnetite formation

On type I corrosion patterns, a thick (5–30 μm) magnetite layer has been observed on the outer part of the DPL (Fig. 6). Its regular morphology all around the artefact could indicate that it was formed during the first stages of the corrosion process. Two hypotheses are proposed for the conditions in which it was formed. First, this layer could have formed during heat oxidation caused by hot forging of the nails. However, during such processes, one expects to form wüstite FeO, magnetite Fe_3O_4 and haematite α-Fe_2O_3[23] Neither wüstite nor haematite has been identified here and this could be because they have undergone a transformation during the burial. The second hypothesis is based on corrosion processes in aqueous media at ambient temperature. Figure 7 presents the Pourbaix diagram based on the modelling of the environmental conditions. It reveals that magnetite is stable in the current environmental conditions recorded on site. This could also have been the case during the initial burial conditions, which would have favoured its formation. Furthermore, the expansive volume caused by the formation of the corrosion products could have caused stresses on the item leading to the cracking of the layer.

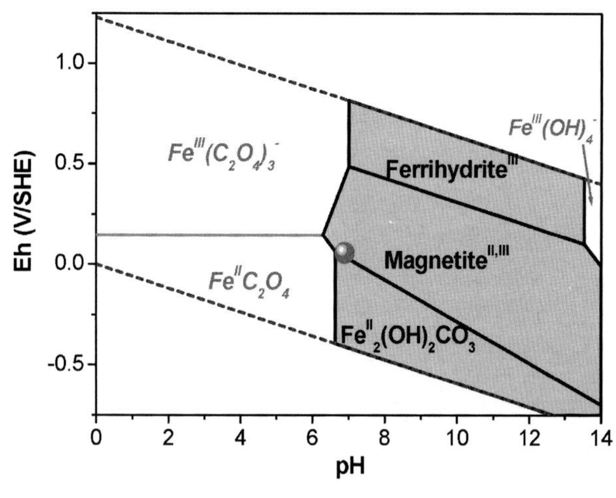

7 Potential–pH diagram of *in situ* conditions [point corresponds to experimental conditions, pH=6·7, Eh=0·134 mV(SHE)]:[24] water composition measured at Glinet;[24] thermodynamic data, Chivot;[25] free enthalpy of chukanovite was assumed to be same as one of malachite[26]

The presence of magnetite nodules inside the DPL could be explained by local changes (redox, pH) during the corrosion process as suggested by the Pourbaix diagram. These changes may have been induced by the corrosion process itself.

Formation of carbonate phases

The potential–pH diagram (Fig. 7) suggests that chukanovite is expected to form under the burial conditions measured. However, despite the fact that this latter phase has been locally identified in the DPL, siderite is the main phase observed on the artefacts. This discrepancy can be explained by the fact that thermodynamic data used for this modelling is for pure phases. Indeed, the occurrence of calcium in the carbonate phases (up to 5 wt-%) should influence the value of their thermodynamic constant and consequently their predominance domain.

Whatever the identified corrosion pattern, when siderite and chukanovite are present, chukanovite is located on the internal zone of the DPL. Thermodynamic data for the carbonate phases[25,27] reveal that calcite is less soluble than siderite, which is less soluble than chukanovite. Consequently, an electrolyte at equilibrium with calcite contains more dissolved carbonate than an electrolyte at the equilibrium with siderite. This makes it possible to suggest the dissolved carbonate content in porewater is higher in the external zone near the calcite layer, than in the internal zone near the siderite layer. Consequently, the carbonate concentration gradient from the outer to the internal part of the DPL imposes following phase order: calcite in the TM siderite in the external zone of the DPL and chukanovite in the internal zone.

Formation of sulphur phases

Iron sulphur phases have not been taken into account in the thermodynamic modelling used in this paper in order to simplify the system. Their occurrence on the external zone of the DPL could be interpreted as being due to the activity of sulphate reducing bacteria.[14] In this hypothesis, sulphur phases could have formed by a recent transformation of carbonate phases. Nevertheless, it is also possible that their formation was induced by local changes of the burial conditions. This hypothesis of the influence of micro-organisms is in good agreement with the calcite formation on the internal part of the TM.[28]

Conclusion

Corrosion diagrams on archaeological artefacts buried in soil have been established using the coupling of microbeam techniques. The main phases are iron(II) carbonates such as siderite $FeCO_3$ and another iron carbonate, that could be chukanovite $Fe_2(OH)_2CO_3$, and magnetite Fe_3O_4. Depending on the location of the phases, three corrosion profiles have been established Using thermodynamic modelling and pore water chemistry to discuss the influence of local variations in pH, redox potential and carbonate concentration on phase stability, reveals that small differences can influence equilibrium values and phase formation.

Acknowledgements

This work is financially supported by the ANDRA and the French Agence Nationale de la Recherche ANR program ARCOR. Moreover, we thank Danielle Arribet-Deroin, the archaeologist who supplied the samples.

References

1. I. D. MacLeod: 'The application of corrosion science to the management of maritime archaeological sites', in 'Marine archaeology: the global perspectives', 411–425; 1995, Dehli, Sundeep Prakashan.
2. H. Matthiesen, D. Gregory, P. Jensen and B. Sørensen: 'Environmental monitoring at a waterlogged site with weapon sacrifices from the Danish Iron age. I: methodology and results from undisturbed conditions', *J. Wetland Archaeol.*, 2004, **4**, 55–74.
3. B. Soerensen and D. Gregory: '*In situ* preservation of artifacts in Nydam Mose', Proc. Int. Conf. on 'Metals conservation', Draguignan-Figanières, France, May 1998, James and James, 94–99.
4. V. Fell: 'Fiskerton: scientific analysis of corrosion layers on archaeological iron artefacts and from experimental iron samples buried for up to 18 months', Centre for Achaeology report no. 65, English Heritage, Portsmouth, UK, 2005.
5. V. Fell and J. Williams: 'Fiskerton, Lincolnshire: analysis of the modern iron analogue samples buried for 30 months', Research Department report series no. 13, English Heritage, Portsmouth, UK, 2007.
6. H. Matthiesen, D. Gregory, B. Sørensen, T. Alstrøm and P. Jensen: 'Monitoring methods in mires and meadows:five years of studies at Nydam mose, Denmark', in 'Preserving archaeological sites *in situ* (PARIS2)', 18–25; 2001, London, Museum of London Archaeology.
7. H. Matthiesen, L. R. Hilbert and D. J. Gregory: 'The occurence and stability of siderite as a corrosion product on archaeological iron from a waterlogged environment', *Stud. Conserv.*, 2003, **48**, (3), 183–194.
8. M. Saheb, D. Neff, P. Dillmann, H. Matthiesen, E. Foy and L. Bellot-Gurlet: 'Multisecular corrosion behaviour of low carbon steel in anoxic soils: characterisation of corrosion system on archaeological artefacts', *Mater. Corros.*, 2008, **59**, 99–105.
9. E. Vega, P. Dillmann, P. Berger and P. Fluzin: 'Species transport in the corrosion products of ferrous archaeological analogues: contribution to the modelling of long-term iron corrosion mechanisms', in 'Corrosion of metallic heritage artefacts: investigation, conservation and prediction for long term behaviour', (ed. P. Dillmann *et al.*), 92–108; 2007, Cambridge, Woodhead Publishing.
10. D. Neff, P. Dillmann, L. Bellot-Gurlet and G. Béranger: 'Corrosion of iron archaeological artefacts in soil: characterisation of the corrosion system', *Corros. Sci.*, 2005, **47**, 515–535.

11. E. Erdös and H. Altorfer: 'Ein dem Malachit ähnliches basisches Eisenkarbonat als Korrosionsprodukt von Stahl', *Werkst. Korros.*, 1976, **27**, 302–312.
12. I. V. Pekov, N. Perchiazzi, S. Merlino, V. N. Kalachev, M. Merlini and A. E. Zadov: 'Chukanovite, $Fe_2(CO_3)(OH)_2$, a new mineral from the weathered iron meteorite Dronino', *Eur. J. Miner.*, 2007, **19**, (2), 891–898.
13. C. Rémazeilles and P. Refait: 'Fe(II) hydroxycarbonate $Fe_2(OH)_2CO_3$ (chukanovite) as iron corrosion product: synthesis and study by Fourier transform infrared spectroscopy', *Polyhedron*, 2009, **28**, (4), 749–756.
14. C. Rémazeilles, D. Neff, M. Saheb, E. Guilminot, K. Tran, J. A. Bourdoiseau, R. Sabot, M. Jeannin, H. Matthiesen, P. Dillmann and P. Refait: 'Microbiologically influenced corrosion of archaeological artefacts; characterisation of iron(II) sulphides by Raman spectroscopy', *J. Raman Spectrosc.*, DOI10.1002/JRS.2717.
15. D. Neff, L. Bellot-Gurlet, P. Dillmann, S. Reguer and L. Legrand: 'Raman imaging of ancient rust scales on archaeological iron artefacts for long term atmospheric corrosion mechanisms study', *J. Raman Spectrosc.*, 2006, **37**, 1228–1237.
16. P. Dillmann and M. L'Héritier: 'Slag inclusion analyses for studying ferrous alloys employed in French medieval buildings: supply of materials and diffusion of smelting processes', *J. Archaeol. Sci.*, 2007, **34**, (11), 1810–1823.
17. M. Saheb, D. Neff, M. Descostes and P. Dillmann: 'Inferences from a corrosion study of Iron archeological artefacts in anoxic soils', *Geochim. Cosmochim. Acta*, 2008, **72**, (12), A818.
18. D. Arribet-Deroin: 'Fondre le fer en gueuses au XVIe siècle. Le haut fourneau de Glinet en pays de Bray (Normandie)', PhD thesis, Paris I Sorbonne, Paris, France, 2001.
19. D. Neff, S. Reguer, L. Bellot-Gurlet, P. Dillmann and R. Bertholon: 'Structural characterization of corrosion products on archaeological iron. An integrated analytical approach to establish corrosion forms', *J. Raman Spectrosc.*, 2004, **35**, 739–745.
20. S. Réguer, D. Neff, L. Bello-Gurlet and P. Dillmann: 'Deterioration of iron archaeological artefacts: μRaman investigation on chlorinated phases', *J. Raman Spectrosc.*, 2007, **38**, 389–397.
21. A. P. Hammersley: ' FIT2D reference manual', ESRF internl report no. EXP/AH/93-02, Grenoble, France, 1993.
22. J. A. Bourdoiseau, M. Jeannin, R. Sabot, C. Rémazeilles and P. Refait: 'Characterisation of mackinawite by Raman spectroscopy: effects of crystallisation, drying and oxidation', *Corros. Sci.*, 2008, **50**, (11), 3247–3255.
23. J. Bénard: 'Oxydation des métaux'; 1962, Paris, Gauthier-Villars.
24. M. Saheb, M. Descostes, D. Neff, H. Matthiesen, A. Michelin and P. Dillmann: 'Iron corrosion in anoxic media: confrontation between ferrous archaeological artefacts and geochemical modelling', *Applied Geochemistry*, In press.
25. J. Chivot: 'Thermodynamique des produits de corrosion: fonctions thermodynamiques, diagrammes de solubilité, diagrammes E-pH des systèmes Fe-H_2O, Fe-CO_2-H_2O, Fe-S-H_2O, Cr-H_2O et Ni-H_2O en fonction de la température', 2004, Châtenay-Malabry, ANDRA.
26. W. Preis and H. Gamsjäger: 'Solid-solute phase equilibria in aqueous solution. XVI. Thermodynamic properties of malachite and azurite–predominance diagrams for the system Cu^{2+}–H_2 O–CO_2', *J. Chem. Thermodyn.*, 2002, **34**, (5), 631–650.
27. D. L. Parkhurst and C. A. J. Appelo: 'User's guide to PHREEQC (version 2) – a computer program for speciation, batch-reaction, one-dimensional transport, and inverse geochemical calculations', 99–4259; 1999, Reston, VA, US Geological Survey.
28. E. Boquet, A. Boronat and A. Ramos-Cormenzana: 'Production of calcite (calcium carbonate) crystals by soil bacteria is a general phenomenon', *Nature*, 1973, **246**, (21–28), 527–529.

Microbiologically influenced corrosion process of archaeological iron nails from the sixteenth century

C. Remazeilles*[1,2], A. Dheilly[2,3], S. Sable[2,3], I. Lanneluc[2,3], D. Neff[4] and P. Refait[1,2]

The presence of Fe and S containing compounds inside rust layers covering iron archaeological nails was suspected but their real nature was not clearly determined. However, this finding suggested that sulphate reducing bacteria (SRB) could be involved in the corrosion processes. A thorough study focused on SRB and FeS compounds potentially present inside the rust layers was achieved on other nails recently excavated. For microbial investigations, the authors used a probe targeting SRB and performed fluorescence *in situ* hybridisation for cells identification. This procedure revealed that SRB were present in all analysed samples. Analysis of rust revealed that $FeCO_3$ was the major component, indicating that the nails remained in anaerobic conditions, but FeS compounds were detected on each sample. Iron sulphides were localised on a few spots in the outer part of rust layers. This shows that the presence and activity of SRB had little influence upon the corrosion system.

Keywords: Microbiologically influenced corrosion, Ferrous archaeomaterials, Anoxic conditions, Sulphate reducing bacteria

This paper is part of a special issue on corrosion of archaeological and heritage artefacts

Introduction

Microbiologically influenced corrosion (MIC) is considered as one of the most severe forms of corrosion of iron and carbon steel and is often characterised by broad pits penetrating deeply the metal. In this phenomenon, bacterial activity modifies heterogeneously the physicochemical properties of the environment close to the metal surface which modifies the anodic and cathodic reactions involved in the corrosion process. The detailed mechanisms of MIC, that can take place once a bacterial biofilm has formed on the metal surface, are not completely understood yet. At the metal/biofilm interface, anoxic conditions can be met and anaerobic microorganisms can develop. Most of works dealing with this subject are focused on the role of anaerobic sulphate reducing bacteria (SRB).[1–4] The metabolic activity of SRB leads to the reduction of sulphates, naturally present in the environment, into sulphides. So sulphides can react with Fe^{2+} ions, which leads to the formation of iron(II) sulphides in the rust layer, otherwise composed by iron hydroxysalts, oxides or oxyhydroxides. However, other micro-organisms can develop on the steel surface, such as sulphur oxidising bacteria and iron oxidising/reducing bacteria, and recent studies showed the existence of a synergistic activity of SRB and iron reducing bacteria.[5] This would imply that biocorrosion involves not only the metallic substrate but also the ferric corrosion products of iron. Consequently, the composition and the morphology of the rust layer covering an object or a structure buried in a biologically active medium may be, at least partially, connected to the various metabolic activities of the whole bacterial population present in its nearest environment and taking part to MIC phenomenon. Archaeological iron nails buried during around 400 years in a soil presenting anoxic conditions, at the site of Glinet (France), have largely been studied. Corrosion layers have been thoroughly described, showing a large amount of carbonated iron(II) corrosion products like siderite ($FeCO_3$) and chukanovite, a Fe(II) hydroxycarbonate with chemical formula of $Fe_2(OH)_2CO_3$.[6–8] A new set of nails was recently excavated for a study focused on the simultaneous detection of iron/sulphur containing compounds by scanning electron microscopy (SEM) and vibrational spectroscopy, and of SRB by fluorescence *in situ* hybridisation (FISH). This method relies on a specific hybridisation of the bacterial rRNA with a 16S rRNA targeted oligonucleotide probe, labelled with a fluorophore. This allows the observation of the bacteria

[1]Laboratoire d'Etude des Matériaux en Milieux Agressifs, EA 3167, Université de La Rochelle, Bâtiment Marie Curie, 17042 La Rochelle Cedex 01, France
[2]Fédération de Recherche en Environnement et Développement Durable, FR CNRS 3097, France
[3]Littoral Environnement et Sociétés, UMR 6250, CNRS-Université de La Rochelle, Bâtiment Marie Curie, 17042 La Rochelle Cedex 01, France
[4]SIS2M/LAPA Laboratoire Pierre Süe, UMR9956, CEA-CNRS, bât. 637, CEA Saclay, 91191 Gif-sur-Yvette, France

*Corresponding author, email celine.remazeilles@univ-lr.fr

Published by Maney on behalf of the Institute
Received 9 December 2009; accepted 11 February 2010
DOI 10.1179/147842210X12659647007167

with a fluorescence microscope. As described in previous works, probe SRB385 is specific for SRB of the *δ proteobacteria* plus several Gram positive SRB within the Clostridia.[9,10] Therefore, this probe was used for the authors' study. Iron sulphides have already been detected on archaeological ferrous artefacts. But studies devoted to the analyses of such compounds are scarce. For instance greigite (Fe_3S_4), mackinawite (FeS), and framboidal pyrite (FeS_2) were identified in the rust layer of several iron objects extracted from a waterlogged soil, at the Iron Age site of Fiskerton (Lincolnshire, UK).[11–13] According to the authors, greigite formed the major component in several samples and was found often associated with siderite. Samples analysed on cross-section revealed that iron sulphides were contained in the outer corrosion layer.[11,12] Other works reporting analyses of artefacts extracted from waterlogged soils confirm that carbonated iron(II) corrosion products are predominant in the rust layer but the possible presence of iron/sulphur containing compounds is not mentioned.[6,7,14] At last a recent work carried out on Roman iron ingots immersed at a depth of 11 m off Les Saintes Maries de la Mer (Provence, France) during two millenniums showed the presence of iron sulphides in the rust layer located in the external part as well.[15] The burial context is different from that of the nails of Glinet but it presents obviously conditions favourable for SRB activity.

In this paper the results obtained on three nails are presented. A specific analytical methodology has been designed, allowing maintaining the micro-organisms active and avoid corrosion resumption and transformation of transient corrosion products, so that the microbiological study and the analysis of the rust layer could be made on a same nail.

Materials and methods

Archaeological nails and sample preparation

The nails come from the archaeological site of Glinet (Normandy, France). This site corresponds to an ancient forge and is dated from the sixteenth century. The nails were excavated in July 2007. In order to keep them unaltered and to preserve the bacterial flora alive from excavation until analysis, they were extracted in their clods of ground and brought quickly to the laboratory. After arrival, they were stored at $-80°C$, until sampling several weeks or months later. It has been demonstrated in previous works[6,7,16–19] that the corrosion system of buried archaeological iron artefacts always presented the same general features. In a cross-section of the system four successive parts can be distinguished:

(i) the metal if the object is not completely corroded
(ii) the corrosion layer adherent to the metal called dense product layer (DPL)
(iii) a more porous and less adherent layer called transformed medium (TM) where corrosion products and tracers of soils are intimately blended
(iv) soil.

Therefore, the original surface of the object more likely corresponds to the DPL/TM interface. For a better consistency between previous and current studies, the same terminology will be used in this text. The objects were systematically used for microbiological analyses first. Their defrosting happened in an anaerobic glove box at 30°C under an atmosphere composed of $90\%N_2$ and $10\%H_2$. Half a nail was scratched to the metal with a scalpel. An external layer, corresponding to a mixture of ground and poorly adherent corrosion products was extracted. This layer has a composition similar to the one of the so called TM. Then the internal layer, composed of dense corrosion products, was sampled. This layer is equivalent to the DPL. After that, the nail was set in a flask full of hexane and closed hermetically in order to shelter it from oxygen once out of the glove box. In the glove box, each sample of DPL and TM was studied separately but the same treatment was applied. Each sample was put in 1 mL phosphate buffered saline (PBS: NaCl, 13×10^{-2} mol L^{-1}; $Na_2HPO_4 10^{-2}$ mol L^{-1}; pH=4·2) and fixed by adding 3 mL fresh paraformaldehyde solution (4% in PBS). The mixture was incubated at 4°C for 15 h and centrifuged (10 min, 8000 g, 4°C). The pellet was resuspended in a 3 mL PBS/ethanol 96% (v/v) solution and stored at $-20°C$ until cells identification experiments. For the subsequent analysis of the corrosion products, the nails were embedded in epoxy resin and cut with a diamond wire saw in order to work on cross-sections. Nujol was used as lubricant to provide an oily protective film for the surface exposed to air during the cutting. To avoid any use of water, the section was polished in hexane with silicon carbide to grade 4000 (grain size of 5 μm).

Characterisation of the rust layers

Analyses started by visual observations of the cross-sections with a stereomicroscope (Leica, M165C). Scanning electron microscopy (SEM-FEI, Quanta 200 FEG/ESEM) coupled with an X-ray microanalysis EDAX system was used for fine observation and determination of the elemental composition of the corrosion products. Observations were performed at high vacuum with an electron acceleration voltage of 20 kV.

In the next step, elemental information was completed by molecular and structural data. Thus, micro-Raman spectroscopy and Fourier transform infrared (FTIR) microspectroscopy were used in order to identify and locate precisely the components of the various strata of the corrosion layers. Raman measurements were carried out at room temperature on a Jobin Yvon High Resolution Raman spectrometer (LabRAM HR) equipped with a microscope (Olympus BX 41) and a Peltier based cooled charge coupled device detector. Excitation was provided by a He–Ne laser at a wavelength of 632·33 nm. Spot diameter was 3 μm. The laser power was reduced by filters to avoid any transformation of the matter by heating. Spectra were recorded with the Labspec acquisition software at a resolution of 2 cm^{-1}. Infrared analysis was performed using a continuum microscope coupled with a Nexus spectrometer (Thermo-Nicolet). The microscope was equipped with one $\times 15$ objective, $\times 10$ oculars and an MCT-A detector which limited the spectral range from 650 to 4000 cm^{-1}. The background was acquired using a gold mirror as a sample. Spectra were obtained with the Omnic acquisition software, at a resolution of 8 cm^{-1} and by averaging at least 128 scans. After acquisition, the specular reflectance spectra were treated by Kramers–Krönig transform. Previous experiments proved that natural Fe(II) hydroxysalts could remain

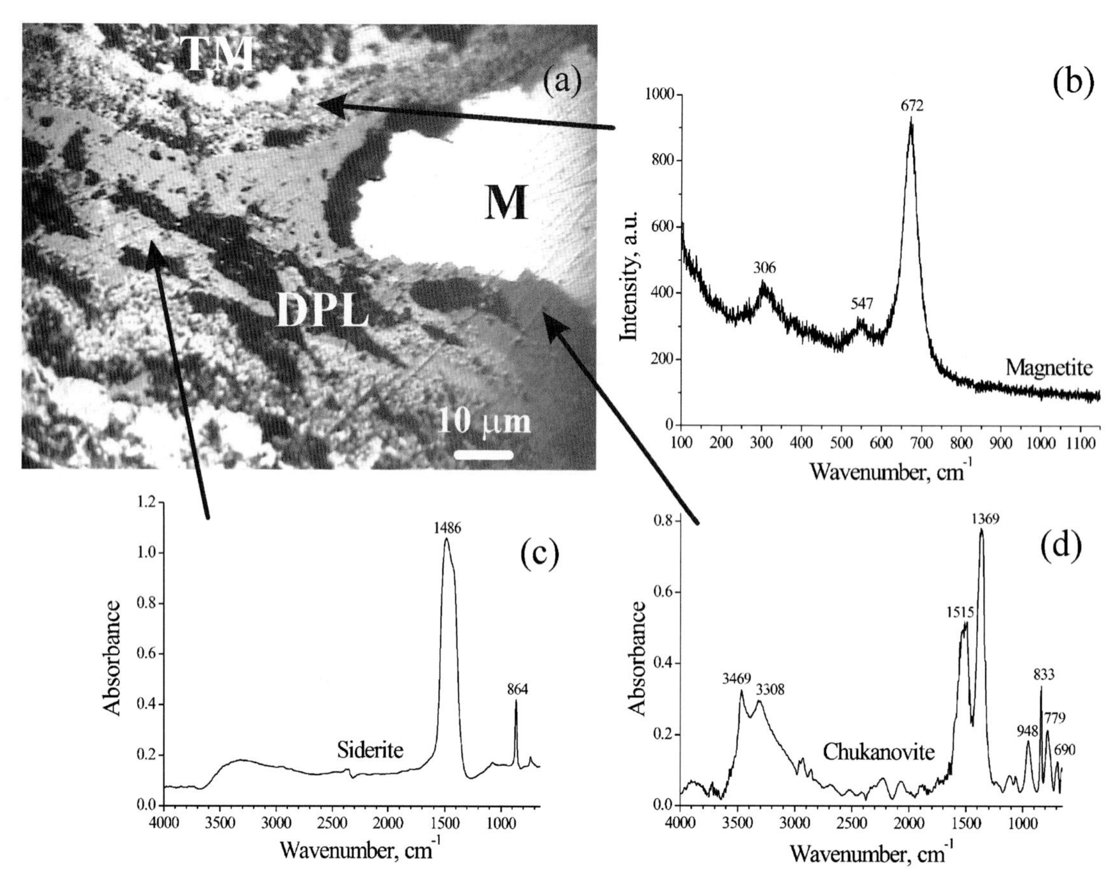

1 ***a*** **optical micrograph of corrosion system of iron nails of Glinet (M, metal; DPL, dense products layer; TM, transformed medium),** ***b*** **micro-Raman spectrum of magnetite and IR spectra of** ***c*** **siderite and** ***d*** **chukanovite present in DPL**

unaffected by an exposure to air during several hours.[6–8] So spectroscopy experiments were carried out without any protection of the samples against oxygen. However, between each experiment the samples were stored in hexane and cross-sections were polished again.

Cells identification by FISH

Fractions of 10 μL mixture stored at $-20^\circ C$ were placed in small wells of a glass slide and allowed to air dry. After dehydration in 50, 80 and 96% ethanol (3 min each), 10 μL hybridisation solution (NaCl, 0·9 mol L^{-1}; 20 mM Tris HCl, 2×10^{-2} mol L^{-1}; pH=7·2; 0·01% sodium dodecyl sulphate (SDS); 20% formamide; 25×10^{-9} g SRB385 probe) were added to each well. SRB385 probe (5′-CGG CGT CGC TGC GTC AGG-3′)[9,10] was Cy3 labelled at the 5′ end. The slide was kept at 46°C for incubation in a humidified Petri dish during 1·5 h. This dish was covered by an opaque film so that the fluorophores were not affected by light. After hybridisation the glass slide was washed at 46°C twice in a saline solution (NaCl, 0·08 mol L^{-1}; Tris HCl, 2×10^{-2} mol L^{-1}; pH=7·2 and 0·01% SDS) during 15 min, then quickly with demineralised water in order to remove all not specifically hybridised probes. At last, the slide was left to dry, sheltered again from light. The samples were visualised with a fluorescence microscope (Zeiss Axioscope 2) emitting a green light at 552 nm, so that the tagged cells appeared red.

To confirm the presence of micro-organisms, di aminido phenyl indol (DAPI, 50×10^{-9} g mL^{-1}) was added to each aliquot. Once attached to DNA, DAPI emits a blue fluorescence at 456 nm under purple light (372 nm). Bacteria were then observed with the fluorescence microscope. A test without probe has also been carried out in order to check that there was no self-fluorescence from our samples.

Results

Cross-section analyses of rust layer

Up to now, several nails coming from the site of Glinet have already been studied. The composition and morphology of the rust layer have been described in previous works.[6,7,16,17] But the case of iron/sulphur containing compounds was not completely clarified. However, the other corrosion products were also analysed in order to verify that the nails considered for the current work presented the same corrosion features as those previously studied.

The DPL of the three analysed nails is characterised by the preponderant occurrence of carbonated phases. Siderite ($FeCO_3$) is the major component detected. But another phase, present here and there close and adherent to the metal surface, could be identified. It is the Fe(II) hydroxycarbonate chukanovite [$Fe_2(OH)_2CO_3$]. Figure 1 shows a stratigraphic view observed by optical microscopy (Fig. 1*a*) and IR spectra of these two carbonated compounds (Fig. 1*c* and *d*). Further from the metallic core, the carbonated layer was often followed by a magnetite border, ~10 μm thick (Fig. 1*b*). After this border, siderite could be detected again by micro-Raman and micro-FTIR spectroscopy but EDAX analyses showed that this siderite contained calcium. The amount of calcium increased with the distance from the metal surface. Quartz and calcite grains were also embedded in this matrix, which implies that this region of the corrosion layer corresponds to the TM. In Table 1 are mentioned

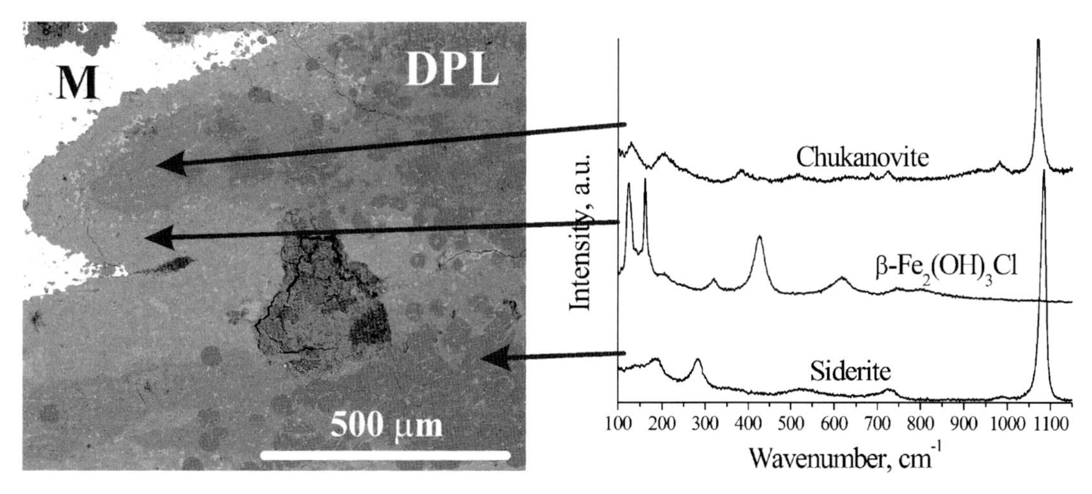

2 Image (SEM) of DPL area composed by Fe(II) corrosion products, and associated micro-Raman spectra

the atomic rates of Fe, Ca and O measured for one sample at several distances from the metal. It shows that the ratio at-%O/(at-%Fe+at-%Ca) is constant and very close to a $Fe_{1-x}Ca_xCO_3$ stoichiometry. This could correspond to a continuous transition from siderite to calcite occurring by a progressive exchange of cations in the crystal structure, and/or to a mixture of phases $(1-x)FeCO_3+xCaCO_3$, richer in $FeCO_3$ close to the nails. It can be noticed that Fe(III) oxyhydroxides like goethite and ferrihydrite have rarely been detected. They were sometimes detected at the interface between the TM and the resin certainly due to an oxidation before embedding the nail. But they were also detected inside the corrosion layers. In this case they were poorly crystallised. This last observation is consistent with possible corrosion resumption during analysis sessions, even if care was taken not to perform too long experiments. Lastly, another compound has been identified in the DPL. The Fe(II) hydroxychloride β-$Fe_2(OH)_3Cl$ was detected here and there on one sample, always between the metal surface and a chukanovite layer. Figure 2 shows the SEM image of a typical area, associated with the micro-Raman spectra corresponding to the various zones. β-$Fe_2(OH)_3Cl$ corresponds to a light zone in contact with the metal. The darkest small nodules are made of siderite. Chukanovite is located between these two phases and is characterised by an intermediate shade of grey. The occurrence of β-$Fe_2(OH)_3Cl$ in the corrosion layers of the nails of Glinet was already observed,[18,19] but the simultaneous presence of these three Fe(II) corrosion products has not been completely interpreted. However, it can be assumed that the formation of the Fe(II) hydroxychloride may have resulted from localised corrosion phenomena due to accumulations of chlorides in cracks or holes of the DPL.

Characterisation of iron sulphides

Iron sulphides could be found in the rust layers of all analysed nails. But sulphur containing zones only appeared locally. Figure 3 refers to an area particularly rich in sulphur. The X-ray map presenting the sulphur distribution (Fig. 3*b*) shows that sulphur rich zones correspond to light zones on the micrograph (Fig. 3*a*), and are mainly located in the TM. On the EDAX spectrum acquired in the TM (Fig. 3*c*), iron and sulphur are observed. But aluminium, silicon and calcium are also largely present, confirming that, in this layer, corrosion products are mixed together with compounds coming from the soil. The same trend was observed for all samples. Micro-Raman spectroscopy proved to be the most efficient tool for the characterisation of iron sulphides because it is suitable for local analysis and because most of the vibration bands of these compounds are active in the far IR range, which cannot be studied with the MCT-A detector of the FTIR microscope. Figure 4 presents two micro-Raman spectra that illustrate what was commonly obtained in the sulphur rich zones. The spectrum in Fig. 4*a* is composed of three main Raman bands at 255, 315 and 355 cm^{-1}. Such a spectrum was already reported and attributed to the so called Fe(III) containing mackinawite.[20] Actually, mackinawite FeS is very sensitive to the oxidising action of O_2 and in the absence of a special care, it will contain Fe(III). It was reported that the structure of mackinawite could withstand up to 20% Fe(III) without noticeable changes in the X-ray diffraction pattern, and so in the crystal structure.[21] The chemical composition of this Fe(III) containing mackinawite would then be $Fe^{II}_{1-3x}Fe^{III}_{2x}S$. However, Raman spectroscopy is sensitive to the presence of these Fe(III) cations in the mackinawite structure and so Fe(III) containing mackinawite can be distinguished from pure $Fe^{II}S$ mackinawite.[20] On the spectra obtained for synthetic Fe(III) containing mackinawite, two bands, at ~310 and ~320 cm^{-1}, could be distinguished.[20] They rather correspond to the broad band obtained here around 315 cm^{-1}. Moreover, the band at 355 cm^{-1} was found much smaller than that seen on the spectrum in Fig. 4*a*. This band may be due to another phase. Actually, the presence of this second

Table 1 Atomic rates and ratio of Fe, Ca and O measured in one sample at several distances from metal

Distance from the metal, μm	Fe, at-%	Ca, at-%	O, at-%	Minor elements, at-%	O/(Fe+Ca)
15	27±2	0±1	70±2	3±1	2·6±0·3
600	23±2	4±1	69±2	4±1	2·6±0·4
1	11±2	14±2	73±2	2±1	2·9±0·4
1·5	1±1	25±2	71±2	3±1	2·7±0·4

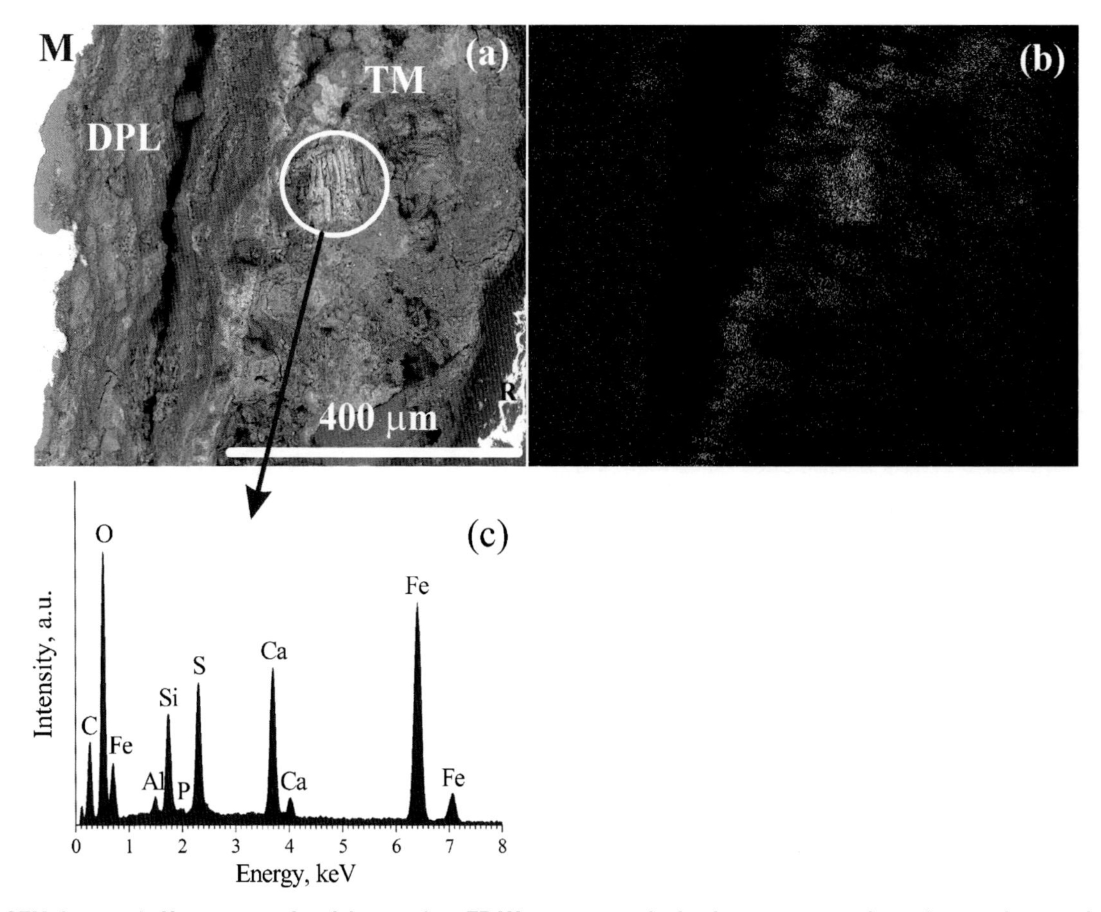

3 ***a*** **SEM image,** ***b*** **X-ray map of sulphur and** ***c*** **EDAX spectrum obtained on cross-section of corrosion products of archaeological iron nail (R, resin)**

phase is clearly revealed by the other spectrum (Fig. 4*b*) that exhibits an intense band at 350 cm^{-1}. To our knowledge, such a spectrum has not been interpreted yet and corresponding spectroscopic data are not available in the literature. However, it may be attributed to Fe_3S_4 (greigite). This is discussed thereafter. The smaller bands at 250, 305 and 320 cm^{-1} can be attributed to Fe(III) containing mackinawite.

Sulphate reducing bacteria detection

First of all, the tests without probes proved to be negative under fluorescence microscope, excluding any potential self-fluorescence from the authors' samples. Di aminido phenyl indol staining tests showed a significant amount of micro-organisms (Fig. 5) in the TM as well as in the DPL. Figure 6 shows the fluorescence observed on samples coming from the two layers of the same nail brought into contact with SRB385 probes. The red colour proves that hybridisation with bacteria happened. This result leads to the conclusion that SRB were present in the corrosion layers of the nails coming from the site of Glinet.

Discussion

The corrosion system of the nails analysed for the current study matches the description already proposed from previous works, indicating that all the nails excavated from anoxic zones of the site of Glinet behave similarly. The main component of the corrosion layer is siderite ($FeCO_3$). But other Fe(II) compounds are present as minor components. Chukanovite [$Fe_2(OH)_2CO_3$] and β-$Fe_2(OH)_3Cl$ were detected at the metal/oxide interface. Their presence is explained by the anoxic conditions of

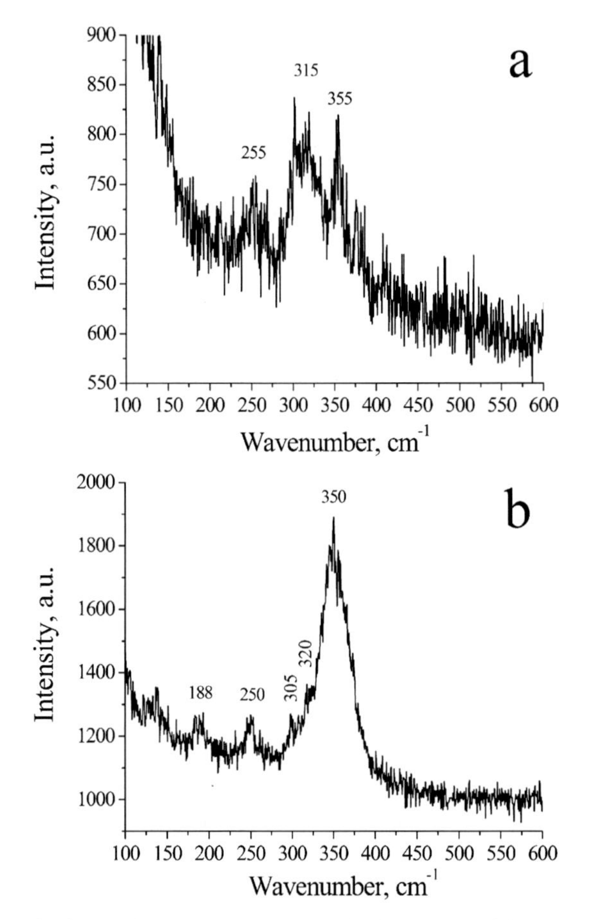

4 Typical micro-Raman spectra obtained in sulphur rich zones ***a*** **Fe(III) containing mackinawite** ***b*** **greigite**

5 Bacteria extracted from *a* TM and *b* DPL scratched around corroded archaeological iron nail, observed under fluorescence microscope after DAPI staining

the soil. As far as the microbiologically activity of this soil is concerned, such conditions are favourable for SRB development. The occurrence of iron sulphides in the corrosion layers of the nails is another hint of the presence, at a given moment, of active SRB that influenced the corrosion system. Two iron sulphides were observed. Among them, Fe(III) containing mackinawite could be clearly identified. This compound is the first step of the oxidation process of mackinawite $Fe^{II}S$. It would appear even in the presence of small amounts of O_2. Raman analysis was carried out without sheltering the samples against oxygen of air, and it can be assumed that some oxidation of mackinawite occurred during the experiments. The other iron sulphide gave a Raman spectrum that, to our knowledge, has not been reported yet. However, this spectrum is mainly composed of a broad Raman band around 350–370 cm^{-1}. Such a band is also present on the spectrum of Fe(III) containing mackinawite and was attributed to the Fe(III)–S stretching mode in the mackinawite structure. So this spectrum may rather correspond to a compound where Fe(III) cations predominate. The more common Fe(III) based iron sulphide is greigite Fe_3S_4, that is $Fe^{II}Fe_2^{III}S_4$. Greigite is known to be an intermediate compound in the oxidation process of mackinawite.[22–24] Moreover, this phase was already identified in the corrosion layer of ferrous archaeological artefacts.[11,12] For all these reasons, it can be assumed that the second phase, obtained together with Fe(III) containing mackinawite, is greigite. The microbiologically study showed the presence of SRB in the entire corrosion layer. No difference seemed to be noticeable between the TM and the DPL whereas iron sulphides were detected only in the TM. The metabolic activity of the SRB in the TM, and of course in the surrounding anoxic soil, may be at the origin of the formation of iron sulphides in the outer part of the rust layers. But in contrast, the identification of SRB in the inner part of the rust layer is not connected to the presence of FeS, as it is for instance the case for steel immersed in seawater.[25] Two explanations can be forwarded. First, SRB may be present in the inner part of the TM, close to the DPL, so that part of the colonies would have been sampled with the DPL. This assumption is consistent with the S mapping in Fig. 3*b* that shows the presence of a border-like S rich region at the inner side of the TM. Second, SRB can only be active if nutrients, necessarily coming from the environment, are available to sustain their metabolic activity. It is thus possible that the SRB present in the DPL are not active due to a lack of nutrients. Anyhow, analysis of the rust layer of the nails demonstrated that iron sulphides were not present close to the metal surface. Such observations are quite consistent with previous works dealing with archaeological iron artefacts extracted either from a waterlogged soil[11,12] or from the bottom of the sea.[15] In consequence it seems that iron sulphides form mainly in the outer part of the corrosion layer, whatever the burial context, provided the environmental conditions are anoxic or at least sufficiently poor in oxygen for SRB colonies to develop. In each case, the corrosion of artefacts was rather uniform and the presence of iron sulphides could never be connected to locally important degradations of the metal. So the presence of SRB in the TM had only little influence on the corrosion processes. To sum up, the iron sulphides biogenerated by SRB are present as minor phases in the outer part of the rust layer, mainly in the TM, and never close to the metal. This implies that at present time sulphide species do not reach the iron surface, which would generate iron sulphides at the iron/DPL interface. Similarly, sulphide species did not reach the iron surface at any intermediate time, which would have generated FeS clusters here and there inside the

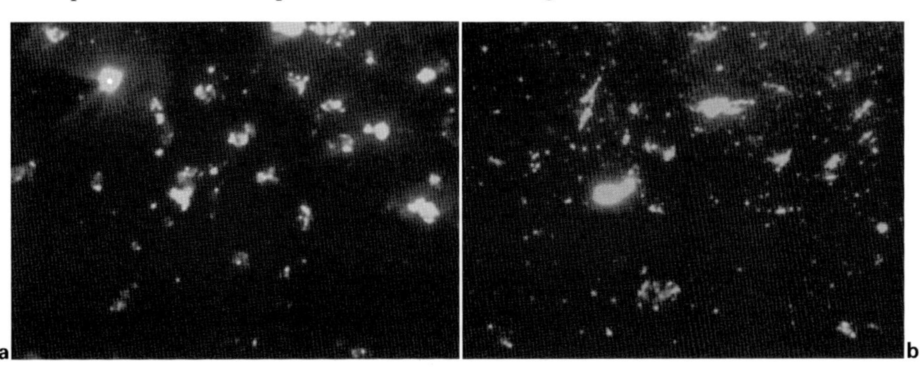

6 Bacteria extracted from *a* TM and *b* DPL scratched around corroded archaeological iron nail, observed under fluorescence microscope after SRB 385 probes hybridisation

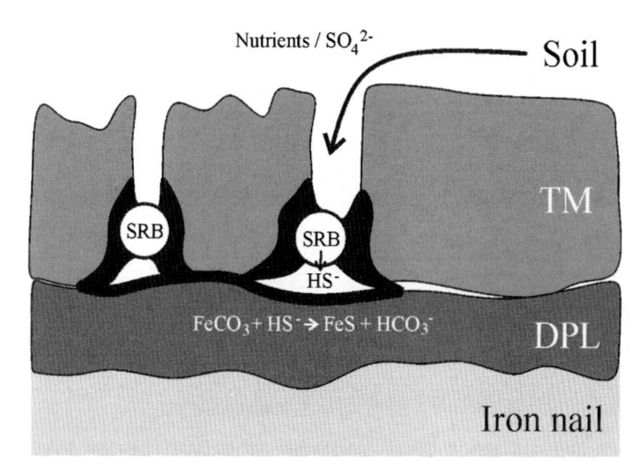

7 Schematic representation of influence of SRB on corrosion system formed by archaeological nails in anoxic soils (TM, transformed medium; DPL, dense products layer)

DPL. So it can be assumed that the iron sulphides detected result from a phenomenon that took place a long time after the formation of the rust layers. A schematic description of this phenomenon induced by SRB is displayed in Fig. 7. Sulphate reducing bacteria present in the anoxic soil surrounding the nails may reach the TM and enter into this layer through large pores or cracks. These pores can insure the continuous transfer of sulphate ions and nutrients from the environment to the SRB colony developing inside the TM. The metabolism of SRB will then produce sulphide species, HS^- being predominant for pH values larger than 7. For simplicity, let us consider only the main iron compound present in the rust layers (TM and DPL), that is siderite. It is known that siderite is more soluble than FeS: $[Fe^{2+}][HCO_3^-]/[H^+]=10^{-0.47}$ and $[Fe^{2+}][HS^-]/[H^+]=10^{-3.00}$.[26,27] So in the presence of HS^-, FeS can precipitate from the solution in equilibrium with $FeCO_3$. The corresponding decrease in the Fe^{2+} concentration induces the progressive dissolution of siderite. The transformation of $FeCO_3$ into FeS should of course take place at the interface between rust layer and pore solution, and should begin in the vicinity of the SRB colony. It seems that the TM does not adhere to the DPL; this is clearly visible in Fig. 4. Actually, this allowed us to separate easily the two layers. So the pores present inside the TM are more likely connected to each other which would explain the formation of the border-like FeS rich zones. The experimental results indicate that the colonies would rather be located in the TM but close to the DPL and Fig. 7 was drawn accordingly. What could be the reason for that? In fact, a too large concentration of sulphide species would be, at last, unfavourable to the development of SRB. Inside the rust layer, the dissolution of $FeCO_3$ and the subsequent precipitation of FeS can maintain the dissolved sulphide species concentration at a low value.

Conclusion

The microbiological and physicochemical study of archaeological iron nails (the sixteenth century) extracted from anoxic soils revealed that interactions between SRB and corrosion products took place. The main component of the rust layer is siderite $FeCO_3$ but Fe(II) hydroxysalts, namely $Fe_2(OH)_2CO_3$ and β-$Fe_2(OH)_3Cl$, were also detected. The use of FISH allowed us to demonstrate that SRB could colonise the external part of the rust layers surrounding the metal, leading to the formation of iron sulphides. This phenomenon did not induce specific damages, even though it took place locally, which could have generated concentration cells. Actually, the H_2S/HS^- species produced by the metabolic activity only induced a partial dissolution of siderite and precipitation of iron sulphides in the external part of the rust layer. This trapping of the sulphide species close to the colonies of SRB more likely prevented their reaction with the metal surface. Therefore, this phenomenon seems to be rather typical of the influence of SRB that interacted with archaeological artefacts already strongly corroded and covered by thick layers ($\sim$500 mm) of corrosion products.

Acknowledgements

This work was supported by the French research national agency (ANR) as part of the 'ARCOR' (ANR-06-BLAN-0313-04) project. The authors would like to thank Danièle Arribet-Deroin for providing access to the archaeological site.

References

1. H. A. Videla and W. G. Characklis: *Int. Biodeterior. Biodegrad.*, 1992, **29**, 195–212.
2. P. Angell: *Curr. Opin. Biotechnol.*, 1999, **10**, 269–272.
3. E. Ilhan-Sungur, N. Cansever and A. Cotuk: *Corros. Sci.*, 2007, **49**, 1097–1109.
4. G. Muyzer and A. J. M. Stams: *Nat. Rev. Microbiol.*, 2008, **6**, 441–454
5. I. B. Beech and J. Sunner: *Curr. Opin. Biotechnol.*, 2004, **5**, 181–186.
6. M. Saheb, D. Neff, P. Dillmann, H. Matthiesen and E. Foy: *J. Nucl. Mater.*, 2008, **379**, 118–123.
7. M. Saheb, D. Neff, P. Dillmann, H. Matthiesen, E. Foy and L. Bellot-Gurlet: *Mater. Corros.*, 2009, **60**, 99–105.
8. C. Rémazeilles and P. Refait: *Polyhedron*, 2009, **28**, 749–756
9. S. Okabe, T. Itoh, H. Satoh and Y. Watanabe: *Appl. Environ. Microbiol.*, 1999, **655**, 107–5116.
10. R. I. Amann, B. J. Binder, R. J. Olson, S. W. Chisholm, R. Devereux and D. A. Stahl: *Appl. Environ. Microbiol.*, 1990, **56**, 1919–1925.
11. V. Fell and M. Ward: Proc. ICOM-CC Metal WG Interim Meet., Draguignan, France, May 1998, ICOM-CC Metal Working Group, 111–115.
12. V. Fell and J. Williams: Proc. ICOM-CC Metal WG Interim Meet., Canberra, Australia, October 2004, ICOM-CC Metal Working Group, 17–27.
13. V. Fell: 'Scientific analysis of corrosion layers on archaeological iron artefacts and from experimental iron samples buried for up to 18 months', Report no. 65/2005, English Heritage Centre for Archaeology, Fort Cumberland, UK, 2005.
14. H. Matthiesen, L. R. Hilbert and D. J. Gregory: *Stud. Conserv.*, 2003, **48**, 183–194.
15. C. Rémazeilles, D. Neff, F. Kergourlay, E. Foy, E. Conforto, E. Guilminot, S. Reguer, P. Refait and P. Dillmann: *Corros. Sci.*, 2009, **51**, 2932–2941.
16. D. Neff, P. Dillmann, M. Descostes and G. Beranger: *Corros. Sci.*, 2006, **48**, 2947.
17. D. Neff, P. Dillmann, L. Bellot-Gurlet and G. Béranger: *Corros. Sci.*, 2005, **47**, 515.
18. S. Reguer, P. Dillmann, F. Mirambet and L. Bellot-Gurlet: *NIMB*, 2005, **240**, 500–504.
19. S. Reguer, P. Dillmann and F. Mirambet: *Corros. Sci.*, 2007, **49**, 2726–2744.
20. J.-A. Bourdoiseau, M. Jeannin, R. Sabot, C. Rémazeilles and P. Refait: *Corros. Sci.*, 2008, **50**, 3247–3255.
21. M. Mullet, S. Boursiquot, M. Abdelmoula, J.-M Génin and J.-J. Ehrhardt: *Geochim. Cosmochim. Acta*, 2002, **66**, 829–836.
22. R. Walker: *Stud. Conserv.*, 2001,**46**, 141–152.
23. R. T. Wilkin and H. L. Barnes: *Geochim. Cosmochim. Acta*, 1996, **60**, 4167–4179.
24. S. Hunger and L. G. Benning: *Geochem. Trans.*, 2007, **8**, 1.
25. S. Pineau, R. Sabot, L. Quillet, M. Jeannin, C. Caplat, I. Dupont-Morral and P. Refait: *Corros. Sci.*, 2008, **50**, 1099–1111.
26. J. Bruno, P. Wersin and W. Stumm: *Geochim. Cosmochim. Acta*, 1992, **56**, 1149–1155.
27. W. Davison, N. Phillips and B. J. Tabner: *Aquat. Sci.*, 1999, **61**, 23–43.

In situ structural characterisation of non-stable phases involved in atmospheric corrosion of ferrous heritage artefacts

E. Burger*[1,2], L. Legrand[3], D. Neff[2], H. Faiz[2,4], S. Perrin[1], V. L'Hostis[5] and P. Dillmann[2,6]

The prediction of very long term corrosion of iron and low alloy steel in atmospheric conditions or in hydraulic binder media is a crucial issue for the conservation and restoration of heritage artefacts. For both media, the typical iron corrosion product layers (CPL) can be described as a matrix of goethite (α-FeOOH) crossed by marblings of reactive phases: maghemite (γ-Fe_2O_3), ferrihydrite ($Fe_5HO_8.4H_2O$), feroxyhyte (δ-FeOOH), etc. The aim of the experiments presented here is to bring new insights on the role that the maghemite could potentially play in the mechanisms of corrosion. For that purpose, electrochemical reductions have been coupled with *in situ* Raman microspectroscopy. These experiments enable the authors to propose a hypothesis of local mechanisms in the specific case of marblings of maghemite connected to the metallic substrate. These local mechanisms could drastically influence the global corrosion rate.

Keywords: Iron corrosion, Electrochemistry, *In situ* Raman microspectroscopy, Maghemite, Corrosion mechanisms

This paper is part of a special issue on corrosion of archaeological and heritage artefacts

Introduction

The prediction of very long term corrosion behaviour of ferrous alloys is crucial for the field of heritage artefacts conservation and restoration.[1] In particular, the understanding of the corrosion mechanisms is necessary to develop a mechanistic model.[2–4] Two different media are concerned in this study: atmospheric corrosion and corrosion in hydraulic (i.e. cementitious) binders. For both media, the corrosion layers after century long periods are several tens to hundreds of micrometres in thickness. Previous studies defined a typical microstructure of long term corrosion product layers.[5–7] It can be described as a matrix of goethite crossed by light marblings mostly made of ferrihydrite ($Fe_5HO_8.4H_2O$), feroxyhyte (δ-FeOOH), magnetite (Fe_3O_4) and maghemite (γ-Fe_2O_3). The nature of the marblings can be influenced by the medium: the reactive phases remaining the most prevalent are ferrihydrite/feroxyhyte/maghemite for atmospheric corrosion, and maghemite/magnetite for century old steels in hydraulic binders.

During the long term corrosion, two cathodic reactions are in competition:[8–13] the reduction of oxygen (reaction (1)) and the reduction of ferric phases in electrochemical contact with the metal core (reaction (2))

$$\frac{1}{2}O_2 + H_2O + 2e^- \rightarrow 2OH^- \quad (1)$$

$$Fe^{3+} + e^- \rightarrow Fe^{2+} \quad (2)$$

It has been proved that the corrosion mechanisms are regulated by successive relative humidity (RH) cycling divided in three stages (wetting, wet stage and drying) and that the predominant reaction could change during this cycling.[8–11] During the wetting stage (corresponding to the growth of the aqueous electrolyte film), the reduction of oxygen is limited by the transport of gaseous oxygen through the pores.[13] The main reaction is thus assumed to be reaction (2). During the wet stage, when all reactive Fe^{3+} species are consumed, only the reaction (1) occurs.

Starting from this basis, it is clear that the corrosion mechanisms are drastically influenced by the nature of these local phases and their physicochemical properties. Thus, the evaluation of their influence on the corrosion processes has become a crucial issue for further mechanistic modelling. Previous electrochemical studies on reference samples show that the most reactive phases of the corrosion system are lepidocrocite, ferrihydrite and feroxyhyte.[14] Furthermore, the reduction of these phases has already been studied by coupling *in situ* technique (XRD, XANES[15] and μ-Raman spectroscopy)[16] to

[1]CEA, DEN, DPC, SCCME, LECA, F-91191 Gif Sur Yvette, France
[2]UMR CEA-CNRS, UMR 3299, IRAMIS, SIS2M, LAPA, F-91191 Gif Sur Yvette, France
[3]LAMBE UMR8587, Université d'Evry/CNRS, Rue du Père Jarland, Evry Cedex 91025, France
[4]Institut Jean Lamour, UMR 7198 CNRS – Nancy-Université, Nancy, France
[5]CEA, DEN, DPC, SCCME, LECBA, F-91191 Gif Sur Yvette, France
[6]IRAMAT, LMC, UMR 5060, CNRS, F-91191 Gif Sur Yvette, France

*Corresponding author, email emilien.burger@cea.fr

Published by Maney on behalf of the Institute
Received 23 December 2009; accepted 1 April 2010
DOI 10.1179/147842210X12710800383729

1 *a* **photography of Saint-Sulpice sample,** *b* **cross-section micrography of whole corrosion system (metal, CPL and transformed medium),** *c* **microphotography enlargement of CPL, with marblings of maghemite and magnetite,** *d* **photography of Amiens sample and** *e* **cross-section micrography of CPL with marblings of maghemite and ferrihydrite**

electrochemical measurements. However, the role of maghemite in the mechanism is not clearly known, whereas this phase is frequently observed in old corrosion product layers (CPL). Figure 1 illustrates two examples of very ancient corroded artefacts: the Amiens cathedral iron chains[7,15] and the Saint-Sulpice church samples aged in hydraulic binder.[5,17]

The aim of this paper is to investigate the electrochemical reduction/reoxidation of maghemite in slightly alkaline medium, and determine the nature of newly formed compounds during the reactions. For this second point, *in situ* techniques are necessary as most of the newly formed compounds rapidly reoxidise under air.[18] For these reasons, *in situ* microRaman spectroscopy analyses were performed in an electrochemical cell specially design for analyses in reflection mode.

Experimental

Materials

NaOH (30%, RP normapur, Prolabo), NaCl (Fluka), N-tris[hydroxymethyl]methl-3-aminopropane-sulphonic acid (TAPS, pH buffering agent with pKa ~8·4; Aldrich, Milwaukee, WI, USA) were used for electrolytes. Magnetite (Alfa Aesar), maghemite (Alfa Aesar) and graphite (Fluka) were used for the working electrode.

Electrochemistry

Raman *in situ* experiments were performed on a specific cell designed to work in reflection mode (Fig. 2). The cell consists in an aperture and a piston. The latter insures both the air tightness of the cell and the control of the electrolyte thickness. The working electrode is composed of a 20 wt-% maghemite–80 wt-% graphite mixture, pressed on a 1 cm^2 stainless steel grid (mass: 15–20 mg, pressure: 5 t cm^{-2} for 1 min). The pellets were 0·8 cm^2 in surface area and their thickness ranged from 200 to 500 µm. A gold wire and an Ag/AgCl (0·1M NaCl) electrode were employed as the counter and reference electrodes. The thickness of electrolyte was controlled at 500 µm. Bubbling of nitrogen (N_2) with a flowrate of 1 L h^{-1} enable the desaeration of the electrolyte.

Two electrochemical reactions were monitored by microRaman spectroscopy: the reduction of maghemite in slightly alkaline pH buffered anoxic solution (pH 9) followed by the reoxidation of the reduced compounds in the same solution aerated. The solution contains chloride (0·1M NaCl) and buffer solution (0·5M TAPS), following the experimental protocol of Antony *et al.*[14] The reduction was achieved by imposing a cathodic current of 25 µA mg^{-1}. The reoxidation of reduced

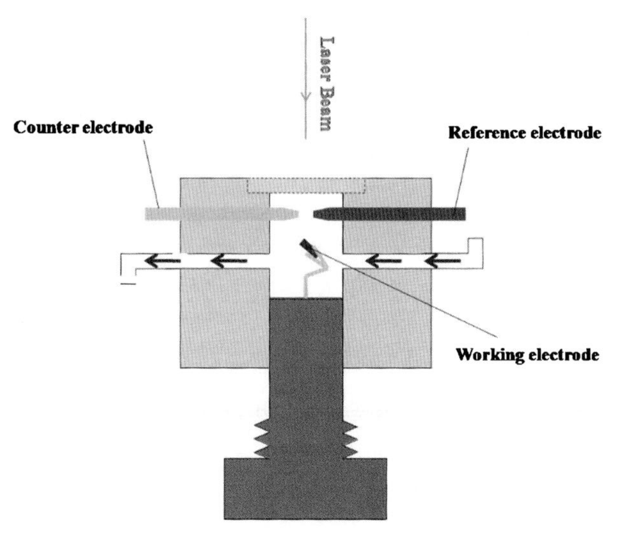

2 Scheme of cell

products was studied by aerating the solution in open circuit.

MicroRaman spectroscopy

In situ microRaman spectroscopy was performed using a Renishaw InVia spectrometer equipped with a frequency doubled Nd:YAG laser at 532 nm. The laser was focused on the sample due to a Leica LM/DM microscope. With the long focus $\times 50d$, the beam diameter is ~ 2 μm and the penetration length is ~ 2 μm. The spectral resolution given by the CCD detector is ~ 2 cm^{-1}. As some iron oxides are highly sensitive to laser radiation,[6] the authors used density filters to control the laser power on the sample under 100 μW.

The evolution of the phase proportions have been quantified from the μ-Raman spectra by using the spectral decomposition CorAtmos program. This program, based on a linear combination of 'pure' reference spectra, fits each experimental spectrum to obtain the phases proportion at each point of the analysed area. Two reference phases have been used: maghemite (γ-Fe_2O_3) and magnetite (Fe_3O_4). Reference μ-Raman spectra of these phases have been previously collected during 500 s with a 20 μW laser power.

Results

Reference Raman spectra revealed that maghemite shows three broad bands around 380, 500 and 700 cm^{-1}, while magnetite shows an intense narrow peak at 667 cm^{-1} with smaller broad bands at 550 and 520 cm^{-1}.[6] Evolution of potential E recorded during both reactions is reported in Fig. 3, while *in situ* Raman spectra during maghemite reduction and reoxidation are respectively reported in Figs. 4 and 5. The small peak at 250 cm^{-1} is related to the chlorate anion from the electrolyte and is not modified during the experiments. This allows asserting that intensities of spectra are directly related to compound concentration.

Reduction of maghemite in pH buffered desaerated solution (pH 9)

Evolution of potential E with time and the associated coulombic charge normalised with respect to one mole Fe (Qt) are recorded during the reduction of maghemite is reported in Fig. 3, and *in situ* Raman spectra is shown in Fig. 4. An increase in the magnetite narrow band (667 cm^{-1}) is observed as well as a decrease in the maghemite bands (380, 500 and 700 cm^{-1}). This clearly confirms a direct reduction of maghemite into magnetite. Moreover, as the whole reduction occurs within the domain of inactivity of the electrolyte [$E > -0{\cdot}9$ V(Ag/AgCl],[14] the reduction of the maghemite and the electrolyte are not taking place concomitantly. This confirms that iron(III) compounds reduction can be the cathodic reaction that balances iron oxidation in the absence of oxygen.

After 104 min, the maghemite seems to have been totally reduced at the surface of the sample (2–3 μm). The evolution of magnetite proportion, quantified by CorrAtmos program (Fig. 6) enabled the authors to estimate a surface reduction rate of 2×10^{-7} mol min^{-1}. As a comparison, the volumic reaction rate, estimated from the coulombic charge Qt, and by assuming that electrons are only used for reaction (3), has been calculated around 4×10^{-8} mol min^{-1}. The reduction is five times slower in the bulk of the pellet than at the analysed surface. This may be due to the contact between the sample and the potentiostat which is locally confined at the surface of the sample. Finally, after 104 min, the surface of the pellets is entirely constituted of magnetite while the bulk is partially reduced ($\sim$40%).

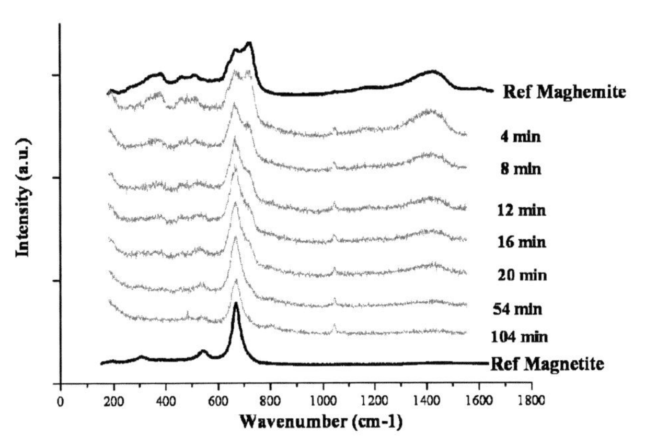

4 *In situ* μ-Raman spectra of intermediate products formed during reduction of maghemite in buffered desaerated solution (pH 9): reduction is achieved by imposing cathodic current of 25 μA mg^{-1}

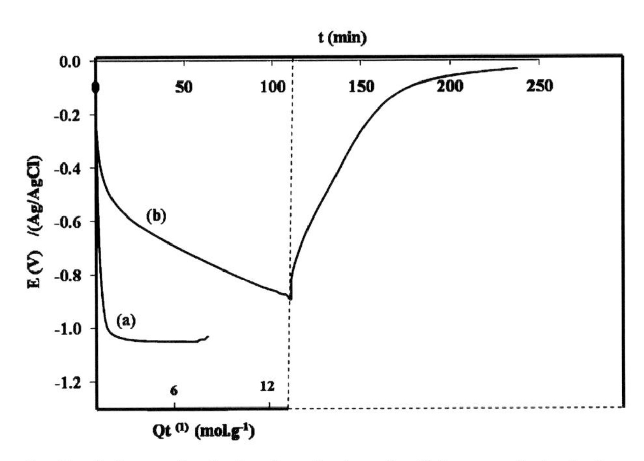

3 Evolution of electrochemical potential recorded during *a* reduction/reoxidation of maghemite and *b* blank graphite electrode, in 0·1M NaCl/0·05M TAPS solution at pH 9, I_c=25 μA mg^{-1}: *E* is given in reference to Ag/AgCl electrode

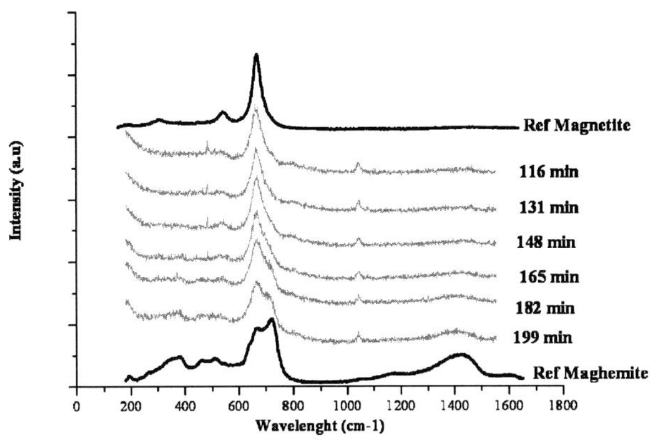

5 *In situ* μ-Raman spectra of intermediate products during reoxidation of reduced compounds in buffered aerated solution (pH 9) in open circuit

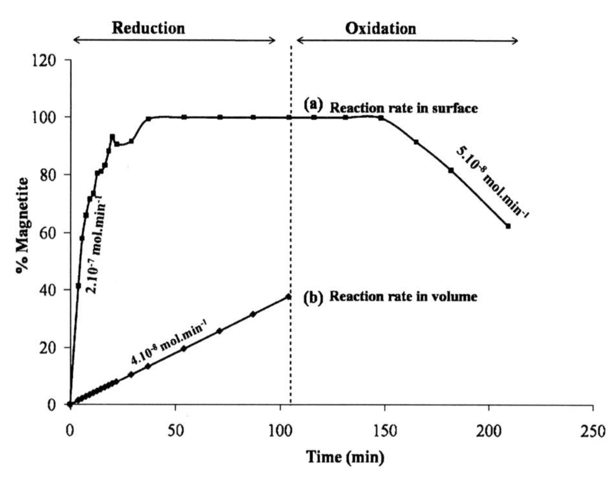

6 Evolution of % magnetite during reduction/reoxidation of maghemite *a* at surface of pellet (estimated from CorrAtmos program) and *b* in bulk of pellet (estimated from E–Q transition)

$$3\gamma\text{-}Fe_2O_3 + H_2O + 2e^- \rightarrow 2Fe_3O_4 + 2OH^- \quad (3)$$

Reoxidation of reduced compounds in aerated pH buffered solution (pH 9) solution

Evolution of potential E recorded during the reduction of maghemite is reported in Fig. 3, and *in situ* Raman spectra in Fig. 5. During the oxidation, a increase in the maghemite bands (380, 500 and 700 cm^{-1}) combined with a decrease in the magnetite one (667 cm^{-1}) is observed. This confirms that the newly formed magnetite is unstable in aerated solution and could potentially regenerate maghemite at the surface of the pellet. However, this reoxidation seems to be very slow. The decrease in magnetite proportion quantified by CorrAtmos program (Fig. 6) enabled the authors to estimate a surface reduction rate of 5×10^{-8} mol min^{-1}. After 100 min, the maghemite regeneration is not totally achieved at the surface of the sample ($\sim$40%).

Interpretation

The experiments presented in this paper enable the authors to propose a hypothesis of corrosion mechanisms (Fig. 7), in the specific case of a marbling of maghemite connected to the metal:

(i) during the wetting stage (Fig. 7*a* and *b*), the iron oxidation is balanced by the reduction of maghemite connected to the metallic substrate. The maghemite reduction leads to the formation of magnetite (reaction (3))

(ii) during the wet stage (Fig. 7*c*), when the marblings of maghemite are totally reduced into magnetite, the cathodic reaction is thus the reduction of oxygen gas dissolved in the electrolyte. The newly formed magnetite conducts the electrons and thus could provoke a decoupling of anodic reaction (metallic substrate oxidation) and cathodic reaction (oxygen reduction) through the conductive marbling. This decoupling drastically influences the corrosion rate, as the transport of oxygen through the whole CPL is rendering unnecessary

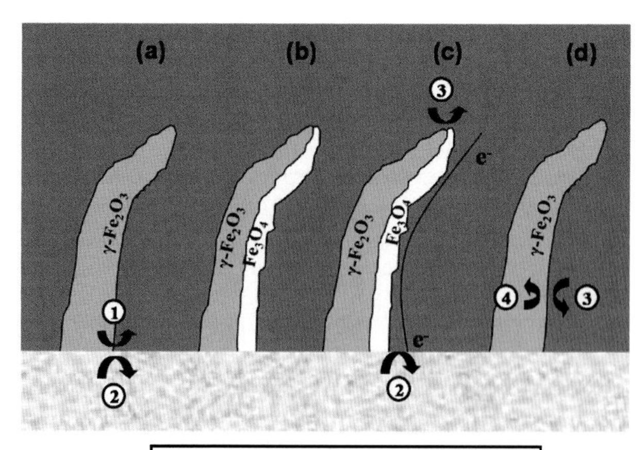

① $3\gamma\text{-}Fe_2O_3 + H_2O + e^- \rightarrow 2Fe_3O_4 + 2OH^-$
② $Fe \rightarrow Fe^{2+} + 2e^-$
③ $\frac{1}{2}O_2 + H_2O + 2e^- \rightarrow 2OH^-$
④ $2Fe_3O_4 + 2OH^- \rightarrow 3\gamma\text{-}Fe_2O_3 + H_2O$

7 Schematic view of local mechanisms due to presence of maghemite marbling in electronic contact with metallic substrate

(iii) during the drying stage (Fig. 7*d*), the concentration of oxygen within the electrolyte film increases as electrolyte layer thickness decreases. In such a medium, the thermodynamic instability of the newly formed magnetite could lead to a progressive regeneration of the initial maghemite during the drying stage.

Finally, the hypotheses proposed here show that marblings of maghemite, when connected to the metallic substrate, could induce local mechanisms which drastically influence the global corrosion rate. Such a mechanism underlines the necessity of a precise microscopic investigation of CPL with a statistical approach before any mechanistic modelling.

The experiments presented in this paper contribute to validating this mechanism and its thermodynamics. First, the reduction of maghemite in surface totally occurs within the domain of inactivity of the electrolyte [$E > -0{\cdot}9$ V(Na/NaCl)],[14] which proves that, for these experimental conditions, maghemite reduces preferentially in regard to the electrolyte. Second, the thermodynamic instability of the newly formed magnetite has been experimentally confirmed. Moreover, the delocalisation of cathodic reaction due to a conductive marblings has been proven by ^{18}O labelling experiments.[15,17]

A complete validation of this mechanism can be brought by proving the compatibility between the kinetic of reactions and the microstructure of naturally formed long term corrosion pattern. In particular, according to the authors' experiments, the magnetite/maghemite ratio within the archaeological CPL strongly depends on the kinetic rate of the maghemite regeneration at the end of the RH cycle. The reaction rate is locally influenced by the physicochemical properties of the CPL. In particular, two parameters are crucial. First, it has been proved that the reduction efficiency of iron oxides is influenced by the pH.[14] In this paper, the reduction/reoxidation of maghemite has been tested at slightly alkaline conditions (pH 9) and the estimated kinetics must be extrapolated to a larger domain of pH. Note that the pH within the CPL is not clearly known yet. Moreover, this latter strongly depends on the ageing media, from cementitious materials to the atmospheric conditions. Second, the hydric properties of the CPL strongly influence its saturation rate.

Conclusion/perspectives

This paper underlines that, at a very local scale, the microstructure of long term corrosion product layer could induce local mechanisms, which drastically influence the global corrosion rate. This paper deals with the specific case of marblings of maghemite in the rust layer, in contact with the metallic substrate. In the first stage, the reduction of maghemite into magnetite could balance the iron oxidation. Then, the newly formed magnetite could provoke a delocalisation of the oxygen reduction at the extremity of the marbling, rendering the transport of oxygen to the metallic substrate locally unnecessary. Moreover, the thermodynamic instability of the newly formed magnetite could lead to a progressive regeneration of the initial maghemite at the end of the RH cycle. This paper contributes to the thermodynamic validation of such mechanism. Moreover, in order to complete the validation of this model, two new approaches are proposed here: a better understanding of the kinetics of the reactions and their extrapolation to various cases (marblings of ferrihydrite, feroxyhyte, etc.) and experimental conditions (pH).

References

1. P. Dillmann, G. Béranger, P. Piccardo and H. Mathiesen: 'Corrosion of metallic heritage artefacts: investigation, conservation and prediction of long term behaviour (EFC 48)'; 2007, Cambridge, Woodhead Publishing.
2. P. Dillmann, F. Mazaudier and S. Hoerle: *Corros. Sci.*, 2004, **46**, 1401–1429.
3. S. Hoerlé, F. Mazaudier, P. Dillmann and G. Santarini: *Corros. Sci.*, 2004, **46**, 1431–1465.
4. W. J. Chitty, P., Dillmann, V., L'Hostis and A. Millard: *Corros. Sci.*, 2008, **50**, (8), 2117–2123.
5. W. J. Chitty, P. Berger, P., Dillmann and V. L'Hostis: *Corros. Sci.*, 2008, **50**, 2117–2123.
6. D. Neff, L. Bellot-Gurlet, P. Dillmann, S. Reguer and L. Legrand: *J. Raman Spectrosc.*, 2006, **37**, 1228–1237.
7. J. Monnier, D. Neff, S. Réguer, P. Dillmann, L. Bellot-Gurlet, E. Leroy, E. Foy, L. Legrand and I. Guillot: *Corros. Sci.*, 2008, **379**, (1–3), 105–111.
8. M. Stratmann and H. Streckel: *Corros. Sci.*, 1990, **30**, (6/7), 681–696.
9. M. Stratmann and H. Streckel: *Corros. Sci.*, 1990, **30**, (6/7), 697–714.
10. M. Stratmann, K. Bohnenkamp and H-J. Engell: *Corros. Sci.*, 1983, **23**, (9), 969–985.
11. M. Stratmann: *Metall. Odlewnictwo*, 1990, **16**, (1), 46–52.
12. C Leygraf and T. E. Graedel: 'Atmospheric corrosion'; 2000, New York, Wiley InterScience.
13. D. Landolt: 'Traité des matériaux: corrosion et chimie de surface des matériaux'; 1993, Lausanne, Presse Polytechniques et Universitaires Romande.
14. H. Antony, L. Legrand, L. Maréchal, S. Perrin, P. Dillmann and A. Chaussé: *Electrochem. Acta*, 2005, **51**, 745–753.
15. J. Monnier: 'Corrosion atmosphérique sous abri d'alliages ferreux historiques, caractérisation du système, mécanismes et apport à la modélisation', PhD thesis, Université de Creteil, Paris, France, 2008.
16. M. C. Bernard and S. Joiret: *Electrochim. Acta*, 2009, **54**, 5199–5205.
17. W. J. Chitty: 'Etude d'analogues archéologiques ferreux corrodés dans les liants aériens et hydrauliques. Application à la prédiction de la corrosion à long terme des armatures métalliques de béton armé', PhD thesis, Université de Compiègne, Compiègne, France, 2006.
18. T. Misawa, K. Asami, K. Hashimoto and S. Shimodaira: *Corros. Sci.*, 1974, **14**, 279–289.

Measuring effectiveness of washing methods for corrosion control of archaeological iron: problems and challenges

D. Watkinson*

The individual chloride content of 116 archaeological iron nails from Romano British and Medieval sites in Wales is reported. The meaning and value of chloride concentration recorded as *weight of chloride in object/object weight* is discussed in relation to reporting the effectiveness of washing methods designed to remove chloride from archaeological iron. This is theoretically compared to the concentration value *weight of chloride in object/metal surface area of object* and the difficulty of quantitatively determining the success of washing methods as stability enhancers is discussed. It is concluded that assessing the impact of residual chloride on post-treatment corrosion of archaeological objects has the potential to offer the most significant guide to treatment success.

Keywords: Iron, Archaeology, Corrosion, Chloride, Treatment, Washing, Assessment

This paper is part of a special issue on corrosion of archaeological and heritage artefacts

Introduction

Preserving archaeological materials as part of cultural heritage is considered important in modern day societies and no greater challenge exists than preventing the post-excavation corrosion of archaeological iron. Corrosion in terrestrial burial environments normally produces objects that have a metal core covered by a dense corrosion product layer (DPL), which is overlaid by a more voluminous and less dense transformed layer containing iron corrosion and soil (TL).[1] The DPL/TL layers range from being quite thin, where they cover substantial iron cores mirroring the original size of the object, to thick layers that overlie thin needle-like iron cores. For some objects, all metallic iron is lost and only DPL remains.

The DPL is heterogeneous, normally comprising a dense $\alpha FeOOH$ matrix containing Fe_2O_3 and Fe_3O_4 strips and it contains microcracks.[1] During burial chlorides are drawn in from soils as counter ions for Fe^{2+} generated at anode sites on the metal surface beneath the DPL layer[2] and they predominantly exist in solution.[1,2] Small pockets of akaganeite ($\beta FeOOH$) have been detected in the DPL and larger amounts at the metal/DPL interface, along with ferrous hydroxychloride [$\beta Fe_2(OH)_3Cl$], which is thought to be a precursor for $\beta FeOOH$ formation.[3,4] $\beta FeOOH$ forms in the presence of chloride, which it adsorbs onto its surface and occludes in its crystal structure.[5–7]

Post-excavation atmospheric corrosion of archaeological iron involves chloride acting as an electrolyte and the generation of FeOOH polymorphs including $\beta FeOOH$.[8–10] This detaches DPL corrosion layers, disfiguring the object. Dry storage of excavated archaeological iron will concentrate chloride at the metal surface and either $FeCl_2.4H_2O$ or $FeCl_2.2H_2O$ may form. A change to damp storage conditions can hydrolyse these compounds to $\beta FeOOH$. Both $FeCl_2.4H_2O$ and $\beta FeOOH$ can corrode iron at low relative humidity.[9,11,12]

Controlling the post-excavation corrosion caused by soluble chloride and chloride bearing corrosion products is essential to prevent break-up of archaeological iron objects. Controlling corrosion normally involves either chloride passivation by humidity control or removal of chlorides followed by controlled storage. Aqueous washing techniques are often employed in attempts to solvate and remove chloride. Chloride removed during treatment is normally quantified,[13–18] but it is unclear what information this supplies regarding the success of a treatment or the post-treatment stability of objects. Historically, once no further chloride was being extracted into the wash solution, it was assumed the iron was free of all chloride, but post-treatment digestion of expendable objects reveals they often have significant chloride residues.[15–17] Studies attempting to determine 'percentage chloride extraction efficiency' for a given treatment with fixed variables such as treatment time, temperature, solution chemistry and concentration, have revealed erratic and unpredictable chloride extraction for many treatment methods.[14–18] Efficiency has been defined as the percentage of the total chloride within an object which was removed by treatment, where total chloride content is determined by object digestion.[14–16] Assessing

School of History and Archaeology, Cardiff University, Cardiff, South Glamorgan CF10 3EU, UK

*Corresponding author, email watkinson@cardiff.ac.uk

Published by Maney on behalf of the Institute
Received 2 January 2010; accepted 9 July 2010
DOI 10.1179/147842210X12754747500801

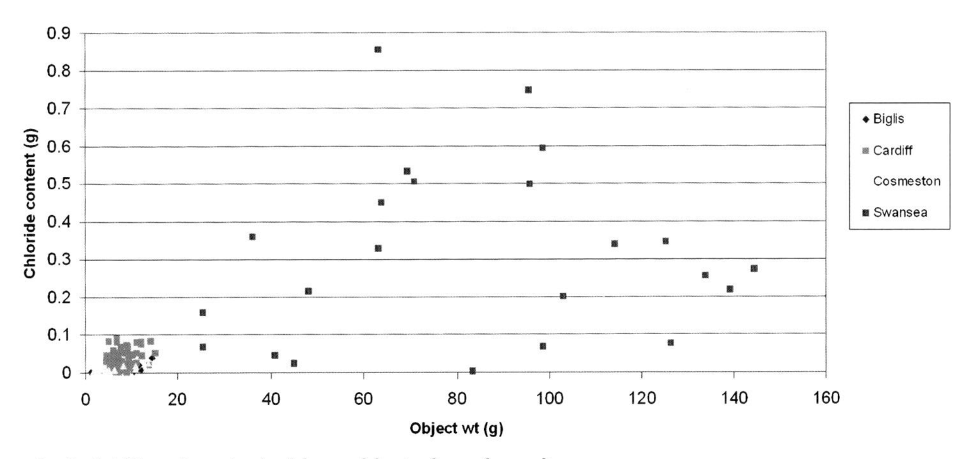

1 Chloride content of 116 archaeological iron objects from four sites

treatment success rarely focused on the quantity of chloride remaining in objects and its potential significance for object stability.

This paper discusses some of the challenges associated with quantifying treatment efficiency for washing methods and the impact of treatments on the corrosion rate of objects by:

(i) reporting the chloride content of 116 archaeological iron objects from terrestrial sites
(ii) examining the relationship between the chloride content of terrestrial archaeological iron, its weight and its surface area
(iii) discussing chloride chemistry in objects and its relevance to chloride extraction
(iv) considering how measuring chloride extracted by treatment contributes to assessing treatment success.

Experimental method

The chloride content of 116 archaeological iron objects that had undergone washing treatments[16,17,19] was determined. They all retained metal cores as determined by X-radiography and were from Welsh archaeological sites. They had been stored in uncontrolled storage environments for several years. Cardiff (70 objects), Newport (23 objects) and Cosmeston (15 objects) are medieval sites and Biglis (8 objects) is Romano British. Cardiff and Swansea objects had been treated individually by a range of aqueous chloride extraction methods (Table 1) reported elsewhere.[16,17] Biglis and Cosmeston had been treated in various sequences by alkaline sulphite, boiling deionised water, Soxhlet (aq.) and pressurised steam.[19] Post-treatment, each object was placed in 5M HNO_3 in a watch glass covered beaker and heated on a water bath daily (not overnight) for up to 3 weeks. Smaller objects digested completely and all corrosion products and some iron core dissolved on the larger objects.

Chloride measurement for neutral aqueous systems was by an EIL 7065 specific ion meter (± 1 mV) using 10 mL samples buffered 1 : 10 with 0·5M HN_4CH_3COO/CH_3COOH. The meter was calibrated with five standard solutions at half day intervals. Modification of this method for different treatments involved preparing the sample according to the treatment procedure used:

(i) Digestion – Cardiff and Swansea: neutralise drop-wise with 0·5M NaOH and wash precipitate free of any chloride into neutralised solution
(ii) $NaOH/Na_2SO_3$: neutralise drop-wise 5M H_2SO_4, heat daily on water bath in a covered beaker for 3 days
(iii) NaOH: neutralise drop-wise 5M H_2SO_4
(iv) $NaSO_3$: heat daily on water bath in a covered beaker for 3 days
(v) Digestion – Biglis and Cosmeston: 2M HNO_3 Fe^{2+} removal in ion exchange column, neutralisation with 2M NaOH.

All methods were tested using known amounts of chloride to establish their reproducibility and the precipitate formed in (i) was digested post-washing and shown to be chloride free.[17] Calculation determined the amount of; chloride removed by treatment; post-treatment residual chloride; total chloride content of each object.

Results

The weight and total chloride content of the 116 archeological iron objects is shown in Fig. 1, with details of objects weighing less than 16 g in Fig. 2.

Table 1 Treatment procedures for archaeological iron*

Treatment	Time, days
0·5M $NaOH/Na_2SO_3$: RT sealed PE container	60
0·5M NaOH deoxygenated: RT in open container in larger box purged with nitrogen	60
Aqueous Soxhlet wash deoxygenated: purged with nitrogen 75–90°C	60–48 cycles per day
Aqueous wash deoxygenated: RT in open container in larger box purged with nitrogen	60
0·5M Na_2SO_3: RT	60
0·5M NaOH: RT	60
Aqueous wash: RT static	60
Soxhlet wash open to atmosphere	60

*RT: room temperature.

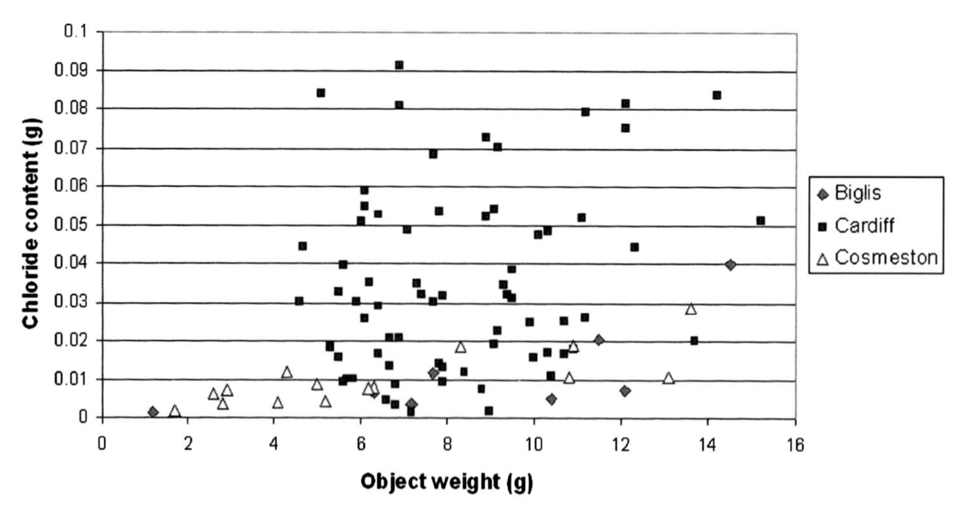

2 Chloride content of 92 archaeological iron objects from Cardiff, Biglis and Cosmeston weight <16 g

The weight of chloride in each object appears to be unpredictable and generally unrelated to object weight (Figs. 1 and 2). All objects >16 g are from the Swansea site, while objects <16 g are from Cardiff, Biglis and Cosmeston. For Biglis and Cosmeston it appears that increasing object weight produces increased chloride content. However, there are too few objects to statistically investigate this apparent trend and neither Swansea nor Cardiff sites produce the same trend for much bigger object populations (Fig. 2). Also, objects of a similar weight from the same site can have greatly differing chloride content (Swansea and Cardiff sites, Figs. 1 and 2). The ceiling content of chloride in objects from Cosmeston and Biglis is lower than that for Cardiff, yet their objects lie within the same weight range as the Cardiff objects (Figs. 1 and 2). As expected the largest amount of chloride is contained in the heaviest objects, yet heavy objects can also contain small amounts of chloride similar to many lighter objects.

Detailed examination shows a 95·6 g iron object contains 0·747 g of chloride, but a 98·6 g object contains 10× less chloride (0·068 g). Similarly, a 6·9 g object contains 0·091 g of chloride and a 6·8 g object contains 11× less chloride (0·008 g). The maximum chloride content is a remarkable 268× the minimum for objects >16 g and 60× for samples <16 g (Figs. 1 and 2). All objects <16 g in weight contain less than 0·1 g of chloride and, while objects >16 g contained up to 0·85 g of chloride, 7 of the 23 objects >16 g fall within the same chloride weight range as samples below 16 g.

Discussion

Chloride content of iron objects

Figures 1 and 2 examine chloride content as a weight value, but chloride associated with archaeological iron is mostly reported as the concentration value; *weight of chloride in object/weight of object*. Plotting this chloride concentration (ppm) for each of the 116 objects against their weight shows how concentration varies with weight of object (Fig. 3). Using a chloride concentration value apparently allows comparison between objects, as lighter objects containing less chloride than heavier objects can have similar chloride concentrations. The chloride concentrations within objects appear to cover a similar range irrespective of object weight; some lighter objects have the highest chloride concentrations, while some heavier objects have the lowest. Comparing sites, Biglis and Cosmeston objects contain lower concentrations of chloride, while objects from the Cardiff and Swansea sites contain similar chloride concentrations that extend over a very wide range of values.

It might be thought that a measure of object stability would be to compare chloride concentrations within objects, using the premise that there will be more free chloride to act as an electrolyte in objects containing high chloride concentrations, irrespective of their weight. This comparison ignores the surface area of metal available for corrosion and thus the concentration of chloride at the metal surface. Calculating chloride concentration per unit surface area of metallic iron is likely to offer better insight into corrosion rate and object stability because theory, measurement and observation suggest that most chloride is held at the metal surface.[4,9,10] This is often localised as chloride nests[4] and their activation in the atmosphere causes objects to break up. For iron objects of the same weight in similar environments, flat thin objects will offer a larger surface area for corrosion to draw in chloride ions, as compared to compact rectangular or cylindrical shaped objects. Consequently, flat objects might be expected to contain more chloride in relation to their weight, mineralise more quickly and potentially disappear entirely in burial contexts when compared to the more compact object morphologies. Unfortunately, calculating the metal surface area on archaeological iron is unrealistic, as it is hidden beneath the DPL and is not flat surface. Only computed tomography with texture analysis can venture to examine this surface without destroying the object by removing all its information retaining corrosion layers. However, experimental studies that link object stability, metal surface area and post-treatment chloride content, could offer useful insight into treatment effectiveness.

A theoretical study of the influence of surface area can be considered. All objects from the sites examined here are either complete nails or nail fragments with roughly generic shapes, being forged from iron bloom to form wrought iron of density ~7·86 g cm^{-3} at 20°C. This

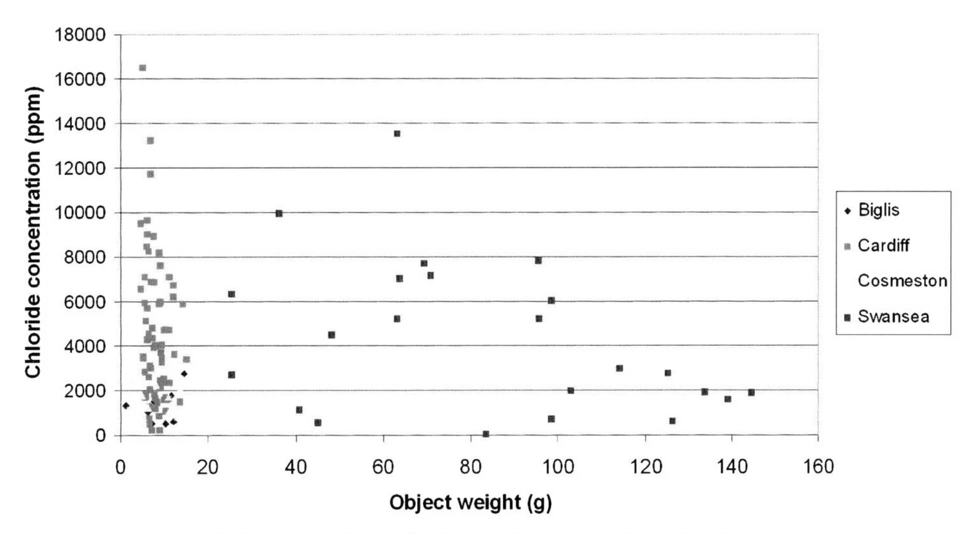

3 Chloride in object (ppm) recorded as weight of chloride in object/weight of object plotted against weight of object

commonality of shape can be used to consider how surface areas vary as a function of weight and how this may influence concentration measurements. For 10 cm long wrought iron bars of varied square cross-section (Table 2), lighter bars offer a larger surface area on which corrosion can occur relative to their weight (Fig. 4). Potentially, with all other factors being equal, they can theoretically pull in more chloride counter ions as a function of their weight. This means that lighter objects could show high concentrations of chloride using *weight chloride in object/weight object* as a concentration determinant. Figure 3 reveals that some of the lightest objects have high chloride concentrations.

Contrary to the expectation of conservators who are responsible for treatment and preservation of archaeological iron, the concentration value *weight of chloride in object/weight object* may not allow comparisons between object stability on the pretext that similar concentrations of chloride indicate similar object stability. Furthermore, no studies have quantitatively linked ppm chloride to object stability in any comparative manner. Theoretically comparing objects of similar generic shape with the same ppm chloride concentration but with differing weights, offers insight into what information *weight of chloride in object/weight object* can provide about object stability. In Table 3 objects of similar morphological shape cited in Table 2 are considered to contain 1000 ppm chloride, calculated as *weight chloride in object/weight object* and their total chloride content is calculated and then related to their surface area.

For this given generic shape, heavier objects have much higher concentrations of chloride per unit surface area than lighter objects, yet their chloride concentration measured as *weight of chloride in object/weight of object* is the same. Since the heavier objects contain higher concentrations of chloride per unit area of metal surface they might be expected corrode more rapidly and lose shape faster than those containing lower concentrations.

Table 2 Surface area, volume and weight of 10 cm bars of wrought iron of differing square cross-sections

Cross-section: side of square, cm	Volume, cm^3	Surface area, cm^2	Weight, g
0·25	0·625	10·130	4·910
0·50	2·500	21	19·65
0·75	5·625	31·250	44·125
1·00	10	42	78·6
1·5	25·50	65·06	200·6
2·0	40	88	314·4

Comparing chloride concentrations in objects as a function of object weight may not provide a relative comparison of stability, but it does offer a crude scale of stability for each object; low chloride concentrations are likely to mean increased object stability. This theoretical comparison between surface areas is undeveloped, as the localisation of chloride in pits and other factors need to be considered to produce a more advanced model. Also, even if small objects corrode more slowly, their localised loss of DPL may soon destroy their shape, whereas large objects may be able to accept some local DPL loss and retain their archaeological value for longer periods. Neither does the model consider the form of chloride present. Also, archaeological objects beginning life with similar generic shapes may be reduced to thin metal cores overlaid by thick DPL layers, which may contribute either significantly or negligibly to object weight according to the object size. Clearly, using the ratio *weight of chloride/weight of object* as a comparator of object stability and treatment success is limited.

Chloride form

Corrosion of archaeological iron is not only influenced by chloride quantity and object morphology, but also by chloride form within the object, as soluble chloride acts as an electrolyte. βFeOOH can corrode iron via its readily soluble surface adsorbed chloride[2,11,12] and some of the chloride from within its crystal structure can be solvated to be available for corrosion processes.[6] While not all the chloride from within βFeOOH can be removed by washing,[6,7] it appears that washing either dramatically reduces or stops its ability to corrode iron.[11,12] Therefore, objects containing similar amounts of chloride, but in differing forms can be expected to corrode at differing rates. No quantified comparisons exist between the corrosivity of soluble chloride and similar amounts of chloride associated with βFeOOH, but an object containing most of its chloride as βFeOOH might reasonably be

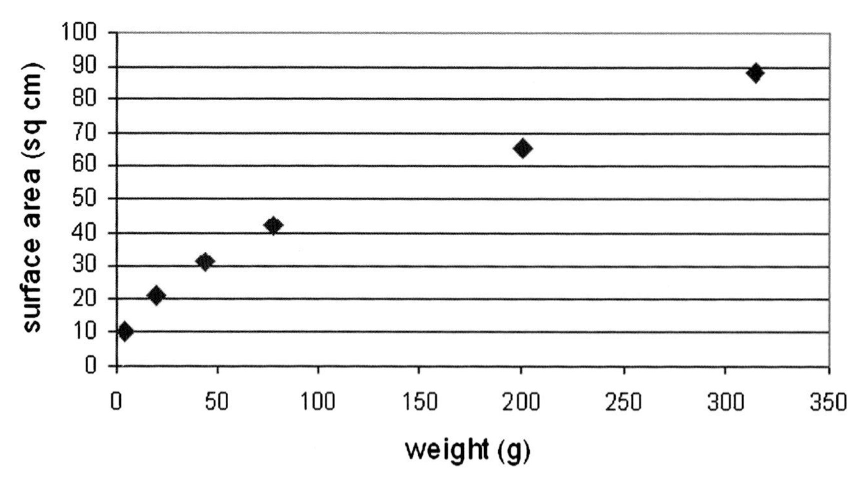

4 Object weight versus object surface area for 10 cm long wrought iron bars with various square cross sections

expected to corrode less quickly than an object containing its chloride entirely in soluble forms, as not all the chloride in βFeOOH is available for corrosion processes.

Paradoxically, although it seems retention of chloride as βFeOOH reflects greater object stability, when compared to objects with chloride in soluble forms, this is not the case when considering the survival of the object as an archaeological record. The driver for controlling the quantity of βFeOOH present on archaeological iron will be the amount of post-excavation corrosion. The quantities of chloride and Fe^{2+} ions present will influence βFeOOH formation. While sufficiently high concentrations of chloride are required to form βFeOOH, low concentrations of Fe^{2+} favour the formation of αFeOOH and γFeOOH even in the presence of high chloride levels.[20] Removing some of the soluble chloride from the corrosion process by 'locking it' into the structure of newly formed βFeOOH requires iron corrosion to generate Fe^{2+} to form the βFeOOH. This iron dissolution weakens the integrity of bonding at the DPL/Fe interface and the resulting βFeOOH growth creates physical pressure on the DPL, destroying the physical integrity of objects which nullifies their heritage value. In light of this destruction, discussing form and relative amount of chloride compounds present on objects may appear irrelevant, but it is relevant to determining the effectiveness of washing methods for increasing the stability of iron.

Assessing treatment success

Chloride quantity and treatment effectiveness

When treating archaeological iron objects using washing methods the chloride content of individual objects is unknown both before and after treatment. Clearly objects are not normally dissolved post-treatment to determine their chloride content and there is no knowledge of how much chloride is removed from a single object since cost, time and object numbers normally dictate that iron small finds are treated in batches, although large objects are treated individually. Knowing collectively the amount of chloride removed from 300 objects does not inform whether any single object contained large or small amounts of chloride. Consider all 116 objects in Fig. 1 treated as one single batch; residual chloride will certainly differ dramatically between objects. Additionally, the chloride forms and their relative quantities are unknown before and after treatment, yet chloride form and quantity will influence post-treatment corrosion rate.

Assessments have shown that treatments produce wide ranges of extraction efficiencies, with deoxygenated alkaline systems generally acknowledged as being the most effective (Table 4).[15,16]

What information can the standard conservation practice of measuring chloride removed from objects provide for persons treating objects? First, it can indicate treatment end points by detecting when no further chloride is being extracted. Second, the quantity of chloride removed can offer a guide as to whether there are large or small amounts of soluble chloride in an object or an object batch. It cannot provide information as to how much chloride objects will retain. Applying best and worst treatment efficiencies reported in Table 4 to the range of objects in Fig. 1 reveals that post-treatment chloride concentration in objects

Table 3 Calculated concentration of chloride per unit surface area for iron objects containing 1000 ppm chloride determined by ratio *weight of chloride in object/weight of object*

Cross-section: side of square, cm	Surface area, cm^2	Weight object, g	Weight chloride in object with concentration of 1000 ppm Cl^-, g	Concentration of chloride per unit surface area, (wt-% Cl^-/SA), μg
0·25	10·130	4·910	0·004910	484
0·50	21·0	19·65	0·019650	936
0·75	31·250	44·125	0·044125	1412
1·00	42·0	78·60	0·078600	1871
1·5	65·06	200·60	0·200600	3083
2·0	88·0	314·40	0·314400	3573

Table 4 Reported treatment efficiencies represented as percentage of total chloride in object extracted by treatment[15,16]

Treatment	Efficiency of chloride extraction as percentage of total chloride in object, %				Number objects in sample
	Average	Std. dev.	Worst	Best	
0·5M Na_2SO_3/NaOH	87	8	68	95	17
0·5M NaOH – aerated	64	16	32	87	16
Soxhlet – water deaerated nitrogen gas	81	12	67	99	10
Water – deaerated static	74	11	62	91	10
Water – aerated static	34	15	12	58	10

can potentially vary significantly. Considering these limitations, it would appear that measuring the effectiveness of treatments is best served, not by applying experimentally determined chloride extraction efficiencies for a given treatment nor by measuring the chloride extracted, but from experimental study of the corrosion capacity of the residual chloride. Variation in residual chloride levels within objects and their individual physical uniqueness and importance as an archaeological record, means that any such assessment should examine the corrosion response of individual objects relative to their post-treatment chloride content. This shifts assessment of treatment success away from the amount of chloride that is removed during treatment, to the amount of chloride remaining in the object and its influence on post-treatment corrosion. To assess this, it is important to consider the form of the chloride before and after treatment.

Chloride form and treatment effectiveness

The relative quantities of chloride counter ions, βFeOOH, $\beta Fe_2(OH)_3Cl$ and $FeCl_24H_2O$ will influence how much chloride washing methods remove from archaeological iron. It is unlikely that all soluble chloride counter ions will diffuse out during aqueous washing, due to object morphology and chloride location at the metal surface beneath dense DPL layers. These pockets of soluble chloride will likely offer the greatest post-treatment corrosion threat. Washing βFeOOH in water and alkali will remove its surface adsorbed chloride[6,7,17] which removes its hygroscopicity and prevents it corroding iron.[6,21] Removal of chloride occluded in tunnels within βFeOOH may not be expected to be complete during washing, but it has been shown that simple aqueous washing for 72 h can remove much of the tunnel located chloride from the βFeOOH without transforming it to other corrosion products[6] and that most chloride lies within the βFeOOH tunnel structure for chloride content below 6 mass-%. Similarly, hot (50°C) and cold deionised water washes lasting 60 h were shown to rapidly remove chloride (41 and 28% respectively) from βFeOOH (4·5 mass-% chloride),[17] although this offers a less complete washing process. Excess OH^- in solution is reported to remove chloride from tunnels within βFeOOH,[22] which gives alkaline washing systems and advantage over other methods. However, there are many synthesis routes for βFeOOH, which can have several forms that hold differing amounts of chloride. All these factors are likely to influence washing effectiveness.

Although washed βFeOOH should offer no corrosion threat,[12,22] the reported metastability of βFeOOH[13] could release its internal chloride sometime in the future. Even though this metastability is in doubt,[12] it would be preferable if washing treatments transformed βFeOOH and released its chloride. While transformation has received limited attention in conservation literature, heated alkaline conditions characteristic of washing treatments have the potential to transform βFeOOH.[5,18,23,24] βFeOOH transformation in KOH at 70°C was influenced by OH^- concentration producing either αFeOOH or αFeOOH and Fe_2O_3.[24] A sample of βFeOOH (4·5 mass-% chloride) boiled in deionised water and left to stand for 4 months was found to have transformed to 18% βFeOOH; 14% αFeOOH; 70% Fe_3O_4.[17] Pressurised NaOH treatment produced full transformation of a βFeOOH/αFeOOH/Fe_3O_4/γFeOOH mix to αFeOOH, Fe_2O_3 and Fe_3O_4.[23] Boiling with water for six days was found to convert βFeOOH to αFe_2O_3[25] and at room temperature βFeOOH in water slowly transformed to αFeOOH.[6] Sodium hydroxide reduced the chloride content of βFeOOH containing 4·5 mass-% chloride to 0·656 mass-% suggesting some transformation had occurred.[17] βFeOOH transformation patterns are varied, unpredictable and merit further study to understand the potential of treatments for removing chloride. It is clear that chloride extraction efficiencies (Table 4) must also record corrosion product transformations and these will influence the quantity and form of residual chloride in objects.

How stable are treated objects?

Discussion here considered the use and meaning of the ratio *weight of chloride in object/weight of object* for the experimental study of treatment efficiency and, by extrapolation, treatment effectiveness. Given the erratic nature of chloride extraction (Table 4) and the large differences in chloride content of objects, it is suggested that determining the effectiveness of treatments lies with the response of treated objects to their environments, rather than focusing on the amount of chloride removed from objects. Quantities of chloride remaining in objects will vary greatly making it essential to focus on how these residues influence iron corrosion. By how much do washing methods improve object stability and how does this relate to optimum storage conditions? There are pertinent resource related questions to answer, such as whether storage humidity can be relaxed from the very low percentages required to stop corrosion in the presence of chloride bearing corrosion products,[11,21] to higher values once objects have been washed. Do washed objects continue to corrode over the same humidity range as unwashed objects, but at slower rates, or is their relative humidity corrosion threshold raised? A more difficult question is how to define object lifespan? The quantity and influence of residual chloride on corrosion, the transformation of βFeOOH, corrosion rates of chloride infested iron and object longevity are currently being investigated at Cardiff University as part of an AHRC funded PhD and an AHRC/EPSRC Science and Heritage large grants award.

An obvious and final question is whether it is worth washing objects, given the treatment unknowns and the resource issues involved? A simple answer to support washing is that lowering chloride levels will reduce corrosion rate. However, this has to be considered in relation to cost benefit, which cannot be assessed without further research. There remain other factors to consider, such as the effect of treatment chemical residues including Na_2SO_3, NaOH and products of their reaction with Fe^{2+} and the atmosphere.[26]

Conclusion

There are wide variations between the chloride content of archaeological objects, even when they are from the same archaeological site. Measuring the chloride removed from archaeological objects in ppm as a function of object weight does not offer a tool for comparing object stability or measuring treatment success, due to morphological, weight and chemical composition differences between objects. New experimental methodologies need to be adopted to examine treatment success. These should focus on the quantity and form of chloride remaining in objects, as well as the corrosion rate of treated objects in various relative humidities. Overall, the vast range of variables involved in treating archaeological iron and the chemical and morphological uniqueness of each object makes quantitative assessment of treatment success a challenging proposition.

References

1. D. Neff, E. Vega, P. Dillmann and M. Descostes: in 'Corrosion of metallic artefacts; investigation, conservation and prediction for long-term behavior', (ed. P. Dillmann *et al.*), European Federation of Corrosion Publications Number 48, 41–76; 2007, Cambridge, CRC Press, Woodhead Publishing.
2. S. Turgoose: in 'Conservation of iron', (ed. R. Clarke and S. Blackshaw), Monograph No. 53, 1–8; 1982, London, National Maritime Museum.
3. F. Zucci, G. Morigi and V. Bertolasi: 'Beta iron oxide hydroxide formation in localised active corrosion of iron artifacts', in 'Corrosion and metal artifacts: a dialogue between conservators, and archaeologists and corrosion scientists', (ed. F. B. Brown), National Bureau of Standards Special Publication 479, 103–105; 1977, NBS Washington DC.
4. S. Reguer, P. Dillmann, F. Mirambet and J. Susini: in 'Corrosion of metallic artefacts; investigation, conservation and prediction for long-term behavior', (ed. P. Dillmann *et al.*), European Federation of Corrosion Publications Number 48, 170–189; 2007, Woodhead Publishing Ltd, CRC Press.
5. R. M. Cornell and U. Schwertmann: 'The iron oxides: structure, properties, reactions occurrences and uses', 2nd edn; 2003, Weinheim, Wiley-VCH.
6. S. Reguer, F. Mirambet, E. Dooryhee, J.-L. Hodeau, P. Dillmann and P. Lagarde: *Corros. Sci.*, 2009, **51**, 2795–2802.
7. K. Stahl, K. Nielsen, J. Jiang, B. Lebech, J. C. Hanson, P. Norby and J. van Lanschot: *Corros. Sci.*, 2003, **45**, 2563–2575.
8. S. Reguer, D. Neff, L. Bellot-Gurlet and P. Dillmann: *J. Raman Spectros.*, 2007, **38**, 389–397.
9. S. Turgoose: *Stud. Conserv.*, 1982, **27**, 97–101.
10. L. S. Selwyn, P. J. Sirois and V. Argyropoulous: *Stud. Conserv.*, 1999, **44**, 217–232.
11. D. E. Watkinson and M. R. T. Lewis: Proc. Int. Conf. on 'Metals conservation', (ed. J. Ashton and D. Hallam), 88–103; 2004, Canberra, National Museum of Australia.
12. D. E. Watkinson and M. R. T. Lewis: in 'Materials issues in art and archaeology VII', (ed. P. B. van Diver *et al.*), 103–114; 2004, Warrendale, PA, Materials Research Society.
13. D. A. Scott and N. I. Seeley: 1987, **32**, 73–76.
14. D. Watkinson: *Stud. Conserv.*, 1983, **29**, 85–90.
15. D. E. Watkinson: in 'Archaeological conservation and its consequences', (ed. A. Roy and P. Smith), International Institute for Conservation, London. 208–212; 1996.
16. D. Watkinson and A. Al Zahrani: *Conservator*, 2008, **31**, 75–86.
17. M. J. Drews, P. de Vivies, N. G. Gonzalez and P. Mardikian: Proc. Int. Conf. on 'Metals conservation', (ed. J. Ashton and D. Hallam), 247–260; 2004, Canberra, National Museum of Australia.
18. A. Al Zahrani: 'Chloride ion removal from archaeological iron and βFeOOH', PhD thesis, Cardiff University, Cardiff, UK, 1999.
19. D. E. Watkinson: 'A comparison of chloride extraction methods for archeological ironwork', Master's dissertation, Cardiff University, Cardiff, UK, 1983.
20. I. Wiesner, B. Schmutzler and G. Eggert: in 'Metal 07: Interim Meeting of the ICOM-CC Metal Working Group', (ed. C. Degrigny *et al.*), Vol. 5, 'Protection of metal artefacts', 110–114; 2007, Amsterdam, Rijksmuseum.
21. D. Watkinson and M. R. T. Lewis: *Stud. Conserv.*, 2005, **50**, 241–252.
22. J. Cai, J. Liu, Z. Gao, A. Navrotsky and S. L. Suib: *Chem. Mater.*, 2001, **13**, (12), 4595–4602.
23. P. de Vivies, D. Cook, M. J. Drews, N. G. Gonzalez, P. Mardikian and J. B. Memet: in 'Metal 07: Interim Meeting of the ICOM-CC Metal Working Group', (ed. C. Degrigny *et al.*), Vol. 5, 'Protection of metal artefacts', 26–30; 2007, Amsterdam, Rijksmuseum.
24. R. M. Cornell and R. Giovanoli: *Clays Clay Miner.*, 1991, **39**, (2), 144–150.
25. J. D. Bernal, D. R. Dasgupta and A. L. Mackay: *Clay Mineral. Bull.*, 1959, **4**, 15–30.
26. M. B. Rimmer and D. Watkinson: Forthcoming in ICOM-CC Metals 2010, Charleston, WV, USA, October 2010.

Influence of corrosion products nature on dechlorination treatment: case of wrought iron archaeological ingots stored 2 years in air before NaOH treatment

F. Kergourlay*[1,2], E. Guilminot[3], D. Neff[1], C. Remazeilles[4], S. Reguer[2], P. Refait[4], F. Mirambet[5], E. Foy[1] and P. Dillmann[1]

Three wrought iron ingots immersed during 2000 years at 12 m deep in Mediterranean Sea were stored after excavation for 2 years without specific protection in air. After that period, two of them were treated by immersion in a NaOH solution, while the third was used to describe the corrosion system resulting from the storage conditions. This characterisation was achieved by a combination of microanalytical techniques. It could be concluded that though ferrous hydroxychloride β-$Fe_2(OH)_3Cl$ was the main Cl containing phase at the time of excavation, akaganeite [β-$FeO_{1-x}(OH)_{1+x}Cl_x$] was the only one present in the rust layers after storage. In order to determine the influence of corrosion products nature on dechlorination treatment, the evolution of a corrosion system composed of both Cl containing phases β-FeOOH and β-$Fe_2(OH)_3Cl$ has been followed during *in situ* NaOH experimental treatment. Specific behaviours of each phase to the dechlorination treatment have been revealed.

Keywords: Storage, Dechlorination treatment, *In situ* treatment cell, Ferrous hydroxychloride, Akaganeite

This paper is part of a special issue on corrosion of archaeological and heritage artefacts

Introduction

After excavation of archaeological artefacts immersed in marine environment, the sudden supply of oxygen combined with the presence of chloride ions trapped within the corrosion product layers increases the decay due to corrosion, which can lead to a rapid destruction of the artefact.[1,2] For that reason, dechlorination treatments have been developed to stabilise artefacts by extracting chloride ions. The different methods applied in conservation laboratories are the chemical immersion in neutral or alkaline solutions,[3] subcritical water,[4,5] immersion coupled with a cathodic polarisation[6] and hydrogen plasma.[7,8] The present study focuses on the immersion in alkaline solution (NaOH) method, most commonly used in restorer workshops. Although this treatment has been applied successfully for several decades for the chloride ions removal, the lack of knowledge about its physicochemical action limits its improvement.[9] One way to reach a better understanding of dechlorination mechanisms occurring during immersion in NaOH solution is to fully characterise the corrosion products at microscopic scale (morphology, localisation of Cl containing phases in the rust layer, etc.). Until now, the corrosion pattern formed before excavation has remained controversial. It was first assumed that solid ferric oxychloride FeOCl was the main Cl containing compound present within the corrosion product layer of marine iron.[3,10,11] Its presence was questioned[12–14] and it was proposed that chloride ions were trapped within the lattice structure of akaganeite β-FeOOH and in pockets of ferrous chloride $FeCl_2$ at the corrosion products/metal interface.[12] In addition, phases such as goethite α-FeOOH, lepidocrocite γ-FeOOH, haematite α-Fe_2O_3 and magnetite Fe_3O_4 have been identified by X-ray diffraction but without any information on their location within the corrosion product layer.[13] Recently, the use of complementary analytical techniques at the microscopic scale[15] has underlined for the first time the presence of the ferrous hydroxychloride β-$Fe_2(OH)_3Cl$, phase already observed on terrestrial archaeological artefacts,[16] as main Cl containing phase of wrought iron artefacts immersed in sea water. The synthesised phase β-$Fe_2(OH)_3Cl$ is known to transform into akaganeite by oxidation mechanisms under specific conditions,[17,18] which would explain the presence of akaganeite on marine iron after

[1]SIS2M/LAPA, CEA/CNRS et IRAMAT LMC UMR5060 CNRS, CEA Saclay, Gif sur Yvette Cedex, France
[2]DIFFABS, Synchrotron SOLEIL, Saint-Aubin, France
[3]Laboratoire ARC'ANTIQUE, Nantes, France
[4]Laboratoire d'Etudes des Matériaux en Milieux Agressifs (LEMMA), Université de La Rochelle, La Rochelle Cedex, France
[5]Centre de Recherche et de Restauration des Musées de France (C2RMF), Paris, France

*Corresponding author, email florian.kergourlay@cea.fr

Published by Maney on behalf of the Institute
Received 25 December 2009; accepted 13 June 2010
DOI 10.1179/147842210X12767807773448

1 Ingot stored for 2 years under ambient atmosphere and detached corrosion products

exposure to oxygen. Although it is commonly assumed in the literature that akaganeite is formed after excavation,[9,12,14] the possibilities for its formation remain open.[16,19]

The aim of this study is first to characterise, by a set of complementary and microbeam analytical tools, the nature and the localisation of the corrosion products formed on marine artefacts kept in air storage. The second aim is to understand the effect of the NaOH dechlorination treatment on the two main Cl containing phases constituting the corrosion layers ferrous hydroxychloride β-$Fe_2(OH)_3Cl$ and akaganeite β-$FeO_{1-x}(OH)_{1+x}Cl_x$ respectively, for freshly and stored marine iron artefacts.

Materials and methods

Archaeological corpus

The studied objects were 2000-year-old iron ingots immersed at 12 m deep and 1·5 miles from the coasts near Les Saintes-Maries-de-la-Mer in Mediterranean Sea. Three ingots were excavated and stored during 2 years in ambient atmosphere without environmental control. During this storage period, a peeling and cracking of the corrosion products at the surface of the ingots was observed (Fig. 1). In order to monitor the evolution of the corrosion layer during a treatment, an ingot was dedicated to the study of the corrosion system before the treatment and the two others were treated by dechlorination. Densities of the ingots were calculated from the ratio of mass to the estimated volume. The obtained values are similar so that a comparison of the chloride ions quantity extracted during the treatment can be made directly from experimental measurements.

Treatment protocol

The two treated ingots were immersed either at room temperature or 50°C in a 0·25M NaOH solution during 39 days. After that period, the chloride ions concentration in the bath was less than 15 ppm, the threshold value for stopping the treatment. The ratio $V_{solution}/V_{ingot}$ was fixed at 15 to fully immerse the ingot and the chloride ion extraction was followed by argentimetry–potentiometry measurements. All of these dechlorination treatments were achieved at the Arc'Antique Laboratory.

Analytical protocol

Following the dechlorination treatment, the ingots were embedded in epoxy resin and transverse sections were cut for analysis. Each transverse section was ground using ethanol with SiC papers (grade 80–4000) and polished with diamond paste (3 and 1 μm).

First, the morphology of the rust layer was determined by optical microscopy and scanning electron microscope (SEM) observations. Then, elementary composition was obtained by energy dispersive spectrometry (EDS), coupled to SEM, with an acceleration voltage of 15 kV. EDS detection was carried out with a Si(Li) detector equipped with a beryllium window allowing quantifying oxygen with an error of 2% and other elements below 0·5%w content with 1% of error. Finally, structural identification of corrosion products was obtained by microRaman spectroscopy (μRS) and X-ray diffraction (XRD). μRS analyses were carried out

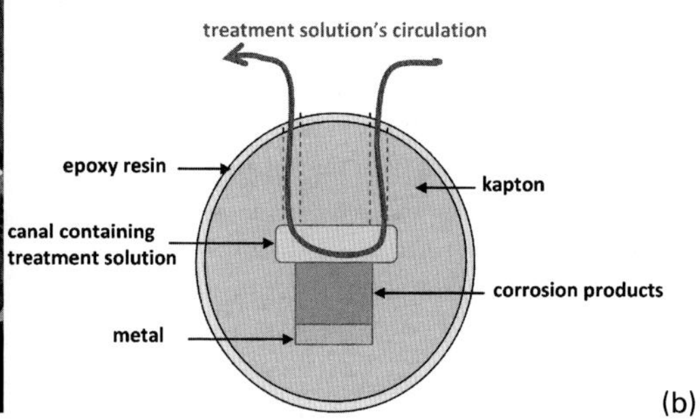

2 *a* XRD set-up and *b* treatment cell (diameter: 1 in.)

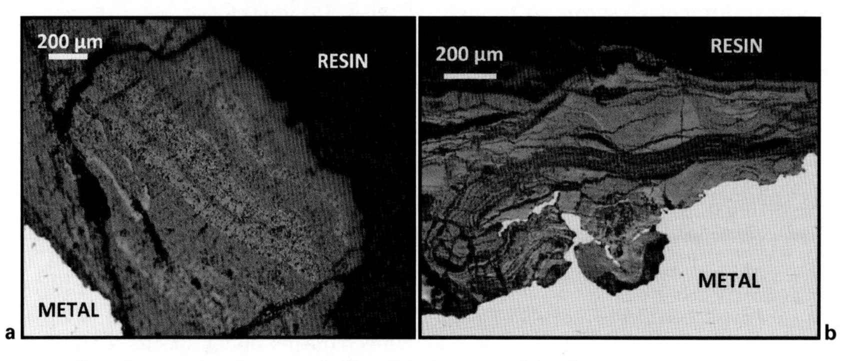

3 Optical microphotographs of corrosion layers *a* before treatment and *b* after treatment

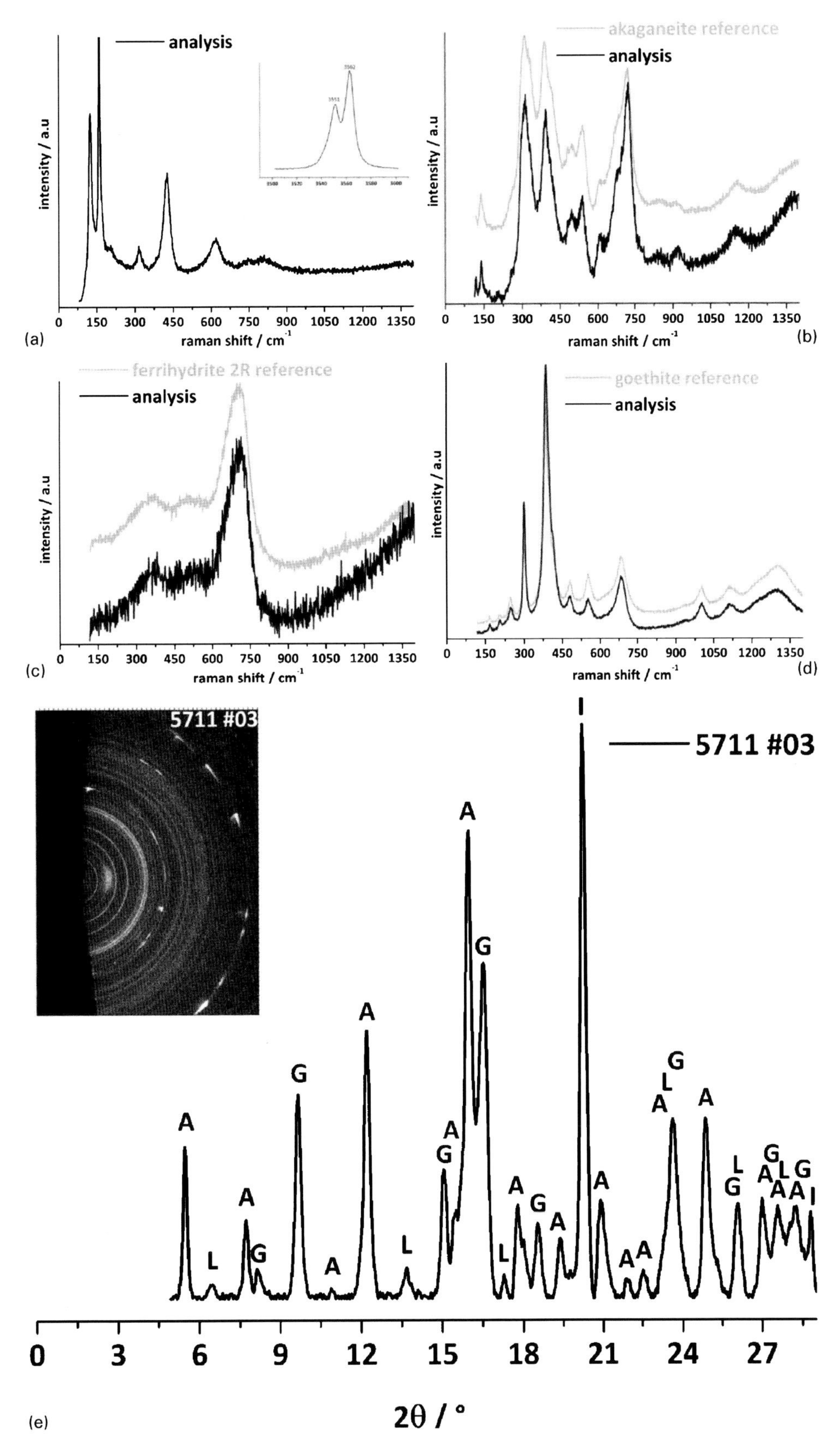

4 Structural investigations by *a–d* μRAMAN and *e* XRD on ingots before treatment: μRAMAN spectra of different phases analysed in corrosion layers were compared to reference spectra except in case of iron hydroxychloride (*a*), identified by comparison to literature;[21,22] on XRD pattern: A: akaganeite β-$FeO_{1-x}(OH)_{1+x}Cl_x$ (JCPDS 080-1770); L: lepidocrocite (JCPDF 044-1415); G: goethite (JCPDF 081-0464); I: iron (JCPDF 006-0696)

using a Renishaw spectrometer (InVia Reflex) with an excitation wavelength at 532 nm. The spot size and the spectral resolution using the ×50 objective were respectively 2 μm and 2 cm^{-1}. Laser power was filtered down to 200 μW to avoid heating and transformation of the corrosion products. The identification of the phases

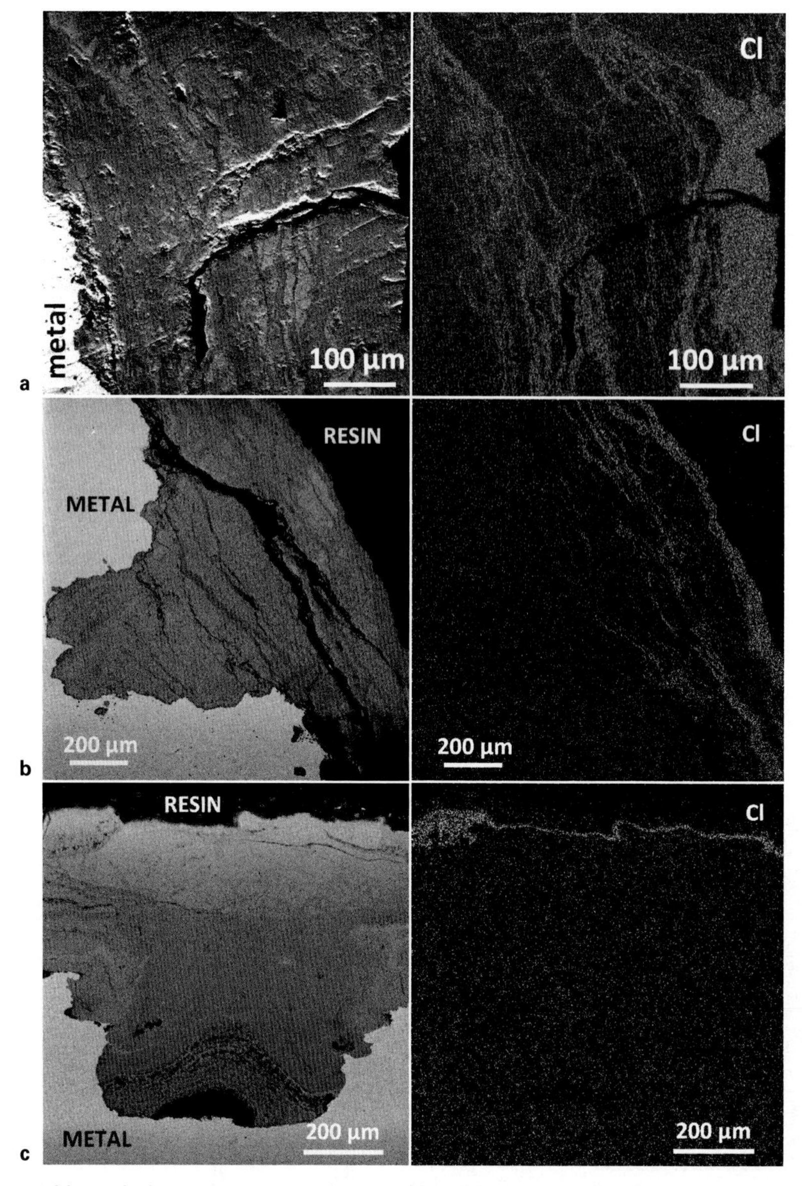

5 EDS–SEM Cl cartographies *a* before treatment, *b* after 0·25M NaOH/room temperature and *c* after 0·25M NaOH/50°C

was established by comparison to spectra reported in the literature.[20–22]

The XRD patterns were collected on a rotating anode generator equipped with a Mo anticathode delivering a monochromatic beam of 30 × 30 µm at 17·48 keV. In reflection mode, the spot size at the sample surface was about 30 × 600 µm. Diffraction patterns were collected using a two-dimensional detector (image plate). Data processing was carried out with the EVA software and the ICDD-JCPDS database.

In situ treatment cell

The aim of this experiment was to monitor the transformation of the phases within the corrosion layer during the first stage of the dechlorination treatment. Therefore, a dedicated cell was designed in order to analyse *in situ* zones within the corrosion layer, while the NaOH solution was passing at the sample surface. For this experiment, a 0·5M NaOH solution was used at ambient temperature. In this case, the ratio $V_{solution}/V_{sample}$ was fixed at 100 in order to avoid a saturation of the treatment solution by chloride ions during the experiment. The treatment cell was made by sampling 1 cm^3 of metal with adherent corrosion products from a transverse section of an ingot. The sample was embedded in resin in a 1 in. diameter mould. A canal was created in the cell to bring the solution near the surface of the corrosion products, using a peristaltic pump. A sheet of Kapton adhesive ensured the sealing of the transverse section to avoid any direct contact of this part of the sample with the solution. The structural evolution of the corrosion products within the layer was followed by XRD in reflection mode (Fig. 2).

Results and discussion

Corrosion products distribution within rust layers of stored artefacts

Before treatment

The corrosion layer has a thickness varying between 200 µm and 2 mm. It displays a marble-like pattern under optical microscope (Fig. 3*a*). The matrix is composed of well crystallised goethite (α-FeOOH) and the marblings are made of poorly crystallised hydrated oxide ferrihydrite ($5Fe_2O_3.9H_2O$) (Fig. 4*c*). In addition, EDS analysis revealed the presence of Cl containing

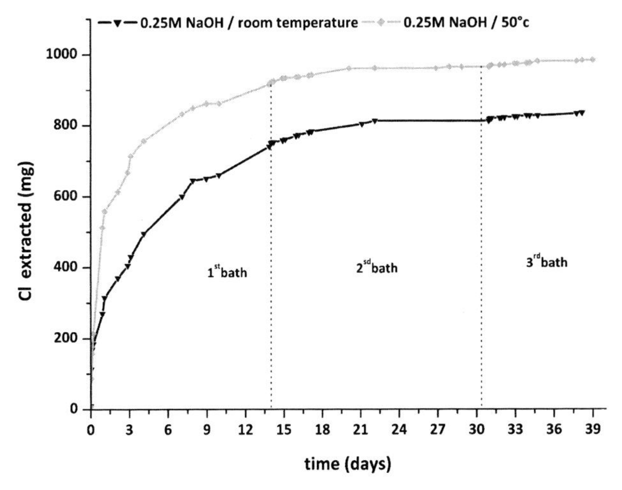

6 Dechlorination curves

veins parallel to ferrihydrite. Their thickness varied from 10 to 100 μm. They were detected indifferently either close to the metal surface or in the outer part of the rust layer (Fig. 5*a*). The veins were identified as akaganeite (Fig. 4*b*) with a Cl content between 6 and 7·5 wt-%. Localised zones of $\sim$500 μm^2 presented chloride contents as high as 8–10 wt-%. The average Cl content in the whole rust layer was $\sim$4wt-%. Finally, lepidocrocite (γ-FeOOH) was sometimes detected by XRD and seemed to correlate with cracks (Fig. 4*e*). It is known that a high oxygen flow would favour the oxidation of Fe(II) containing compounds into lepidocrocite.[23] So it can be assumed that the cracks observed together with lepidocrocite constituted a preferential access for O_2 to the reactive phases present within the rust layer.

After treatment

As expected and already observed in previous studies,[3,24] extraction of chloride ions is favoured by a temperature increase. The amount of chloride ions extracted is increased by 15% at 50°C compared to at room temperature (Fig. 6). After this treatment, the corrosion product layers retained approximately the same thickness. The marblings were still observed under optical microscope. They corresponded to poorly crystallised ferrihydrite (Fig. 3*b*). Cl containing veins, with a thickness of 10–20 μm, were detected by EDS analyses only in the outer part of the rust layer (Fig. 5*b* and *c*). Akaganeite with chloride content between 4 and 6 wt-% was identified in these veins. The average Cl content in the whole rust layer for both ingots were close to 2 wt-% after the treatment. Despite the low Cl concentration in the dechlorination solution at the end of the treatment, the presence of akaganeite in the corrosion products showed that the dechlorination was not complete. Surprisingly, it seems that chloride ions were first extracted from the inner parts of the rust layer (Fig. 5), while an external chlorine depletion would be expected from diffusion processes. This unexpected chlorine distribution could be related to the different classes of chloride ions within the artefacts, adsorbed and bound Cl^-. In fact, the first class of ions is not strongly linked to the structure and can be extracted easily. Nevertheless, this work will be pursued to better understand the location of chlorinated phases in the corrosion layer after treatment.

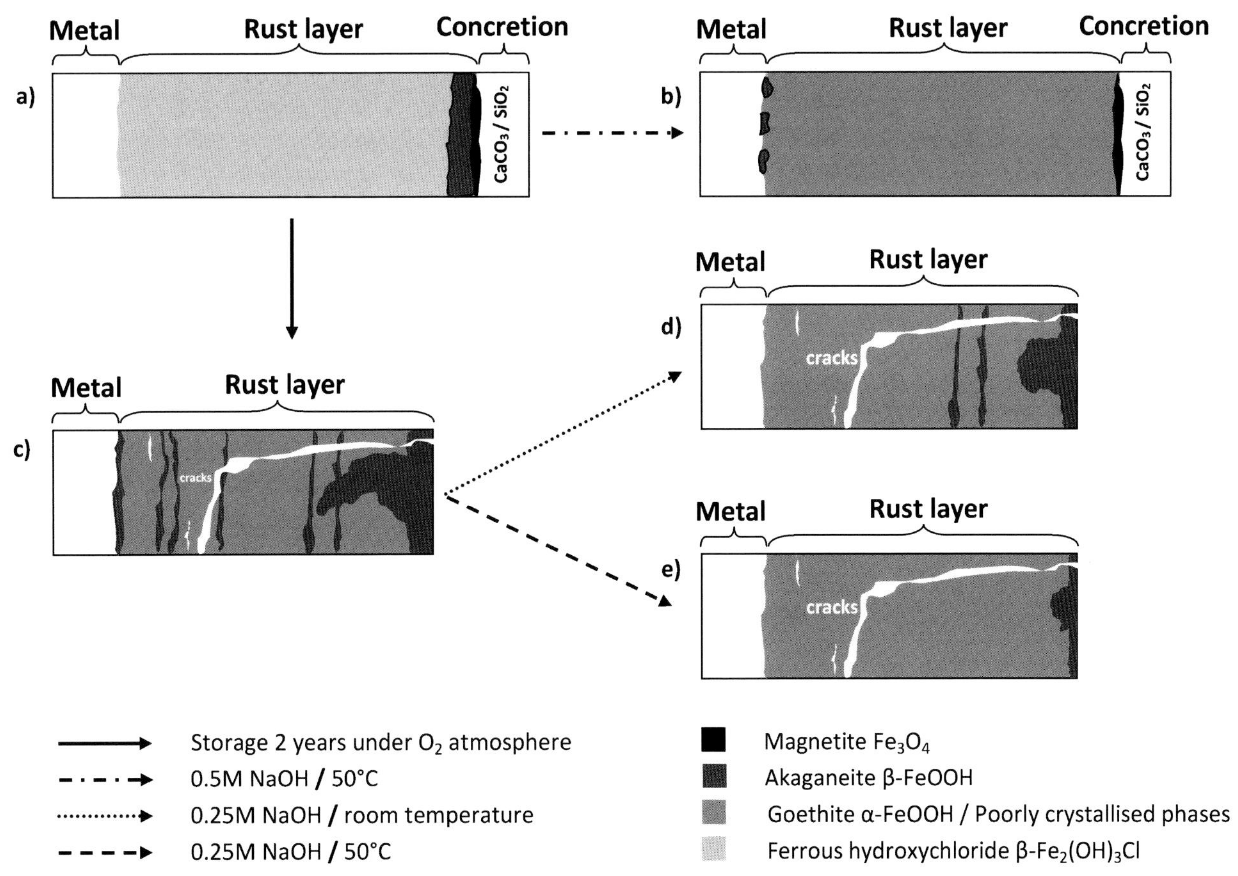

a freshly excavated and stored for a few months in tap water;[13] *b* treated for 42 days; *c* stored for 2 years under O_2 atmosphere; *d* treated for 39 days; *e* treated for 39 days

7 Diagrams of corrosion product layers

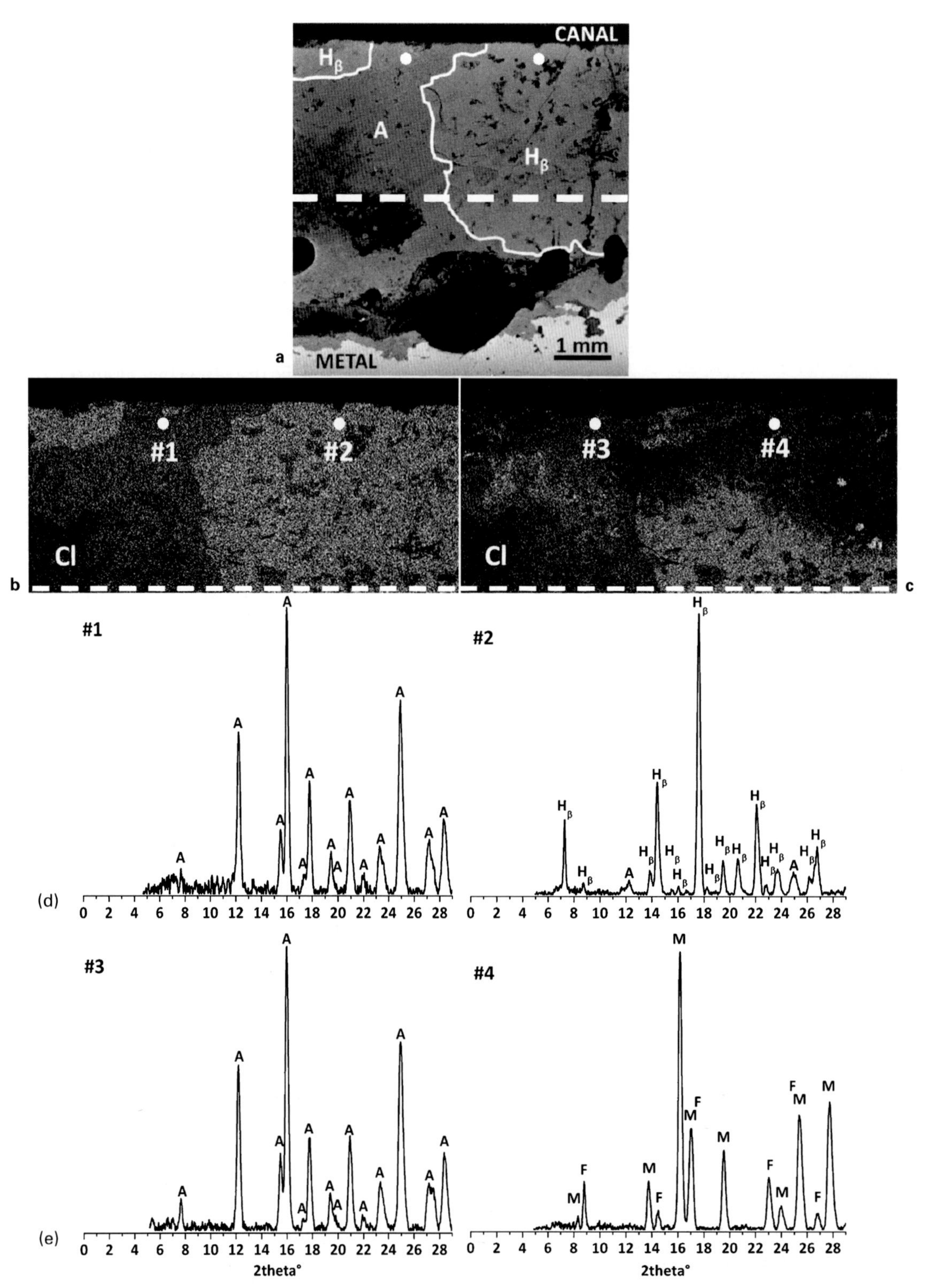

8 ***a*** **sample in treatment cell [H_β: ferrous hydroxychloride β-$Fe_2(OH)_3Cl$; A: akaganeite β-$FeO_{1-x}(OH)_{1+x}Cl_x$; white points correspond to diffracted areas],** ***b*** **Cl X-ray map before treatment and** ***c*** **Cl X-ray map after 3 h of treatment. Diffraction data** ***d*** **before treatment and** ***e*** **after 3 h of NaOH treatment [H_β: ferrous hydroxychloride β-$Fe_2(OH)_3Cl$ (JCPDS 00-034-0199); A: akaganeite β-$FeO_{1-x}(OH)_{1+x}Cl_x$ (JCPDS 01-080-1770); M: magnetite Fe_3O_4 (JCPDS 00-019-0629); F: ferrous hydroxide $Fe(OH)_2$ (JCPDS 00-013-0089)]**

Schematic diagrams of the corrosion product layers can be proposed (Fig. 7*c–e*) and compared to those typical of freshly excavated ingots (Fig. 7*a* and *b*). The main effect of the storage in air is the oxidation of the rather homogeneous layer of β-$Fe_2(OH)_3Cl$[22] into a heterogeneous layer containing veins of akaganeite and marbles of ferrihydrite parallel to the metal/corrosion products interface in a goethite matrix. Dechlorination treatments led to a complete transformation of β-$Fe_2(OH)_3Cl$ into Fe(III) oxyhydroxides; however, they failed to transform all the akaganeite.

In conclusion, the characterisation of the ingots corrosion before and after treatment and with or without storage period, underlines different transformation paths for the corrosion products that have to be understood thoroughly. In particular, the behaviours of the two main Cl containing phases, β-$Fe_2(OH)_3Cl$ and akaganeite, with exposure to the treatment solution, had to be specified.

Therefore, experiments were carried out in a cell specially designed for dechlorination treatments to monitor the transformations of the Cl containing phases *in situ* during the first steps of a treatment.

In situ treatment cell experiment

It was necessary to select a sample with a rust layer containing both Cl containing phases and observable in transverse section (Fig. 4*a* and *b*). An EDS chlorine mapping of the upper part of the corrosion products was collected before the start of the treatment (Fig. 8*b*). Micro-RS identified akaganeite on the left part and β-$Fe_2(OH)_3Cl$ on the right part of the sample. Akaganeite had an average chlorine content between 6 and 9 wt-% and β-$Fe_2(OH)_3Cl$ between 18 and 22 wt-%.

Diffraction data confirmed the presence of the two Cl containing phases in the sample before the start of the treatment (Fig. 8*d*).

Two areas, one for each Cl containing phase, were analysed at the same outer surface distance in order to ensure that the penetration front of the solution had reached both phases (Fig. 8*a*). After 3 h treatment in 0·5M NaOH, on the right part of the sample at a distance of 1–1·5 mm, the transformation of β-$Fe_2(OH)_3Cl$ into magnetite (Fe_3O_4) and ferrous hydroxide ($Fe(OH)_2$) was noted. Where $Fe(OH)_2$ had been identified, the content of chlorine was less than 1 wt-%. In contrast, despite a decrease of nearly 25 wt-% in the average content of chloride in the analysed zone (8 wt-% before treatment and 6 wt-% after 6 h of treatment), the area consisting of akaganeite before treatment remained unchanged. It is likely that only the chloride ions adsorbed on the surface of the akaganeite grains were extracted. Such observations suggested that the removal of the chloride ions trapped within the crystal structure of akaganeite required a deeper and longer treatment, as already mentioned.[25]

Conclusions

This study clearly confirms that the dechlorination treatment by NaOH solution is less efficient for corrosion layers consisting of akaganeite than layers consisting of ferrous hydroxychloride. Akaganeite is the main Cl containing phase present in corrosion layers of artefacts stored in air during several months, while ferrous hydroxychloride dominates within the corrosion layers of immersed artefacts. This could explain the difference in efficiency of dechlorination treatment observed by restorers between artefacts directly treated after excavation or storage in desaerated mediums and those exposed to air for a long period of time.

The next stage of this study is to extend this methodology to the alkaline sulphite and the electrochemical treatments to determine the efficiency of the dechlorination treatment currently used in restorers' workshops.

Acknowledgements

The authors would like to thank the archaeologists, conservators and restorers without whom this study would never been carried out and L. Robinet for his rereading.

References

1. S. Turgoose: *Stud. Conserv.*, 1982, **27**, 97–101.
2. L. S. Selwyn, P. J. Sirois and V. Argyropoulos: *Stud. Conserv.*, 1999, **44**, 217–232.
3. N. A. North and C. Pearson: *Stud. Conserv.*, 1978, **2**, 174–186.
4. M. J. Drews, P. de Viviés, N. G. González and P. Mardikian: Proc. Int. Conf. on 'Metals conservation', Canberra, ACT, Australia, October 2004, National Museum of Australia, 247–260.
5. N. Gonzalez, C. Cook, P. de Viviés, M. Drews, and P. Mardikian Proc. Interim Meet. ICOM-CC Metal WG, Amsterdam, The Netherlands, September 2007, ICOM-CC Metal WG, Vol. 3, 32–37.
6. W. Carlin, D. Keith and J. Rodriguez: *Stud. Conserv.*, 2001, **46**, 68–76.
7. P. Arnoult-Pernot, C. Forrieres, H. Michel and B. Weber: *Stud. Conserv.*, 1994, **39**, 232–240.
8. M. J. de Graaf, R. J. Severens, L. J. van Ijzendoorn, F. Munnik, H. J. M. Meijers, H. Kars, M. C. M. van de Sanden and D. C. Schram: *Surf. Coat. Technol.*, 1995, **74–75**, 351–354.
9. L. S. Selwyn: Proc. Interim Meet. ICOM-CC Metal WG, Canberra, ACT, Australia, October 2004, National Museum of Australia, 294–306.
10. N. A. North and C. Pearson: Proc. ICOM Committee for Conservation 4th Triennal Meet., Venice, Italy, October 1975, ICOM, 1–14.
11. N. A. North and C. Pearson: *Stud. Conserv.*, 1977, **22**, 146–157.
12. M. R. Gilberg and N. J. Seeley: *Stud. Conserv.*, 1981, **26**, 50–56.
13. J. Argo: *Stud. Conserv.*, 1981, **26**, 42–44.
14. N. A. North: *Stud. Conserv.*, 1982, **27**, 75–83.
15. C. Rémazeilles, D. Neff, F. Kergourlay, E. Foy, E. Conforto, E. Guilminot, S. Reguer, P. Refait and P. Dillmann: *Corros. Sci.*, 2009, **51**, 2932–2941.
16. S. Réguer, P. Dillmann and F. Mirambet: *Corros. Sci.*, 2007, **49**, 2726–2744.
17. P. Refait and J.-M. R. Genin: *Corros. Sci.*, 1997, **39**, 539–553.
18. C. Rémazeilles and Ph. Refait: *Corros. Sci.*, 2007, **49**, 844–857.
19. D. Watkinson and A. Al-Zahrani: *Conservator*, 2008, **31**, 75–86.
20. D. Neff, L. Bellot-Gurlet, Ph. Dillmann, S. Reguer and L. Legrand: *J. Raman Spectrosc.*, 2006, **37**, 1228–1237.
21. S. Réguer, D. Neff, L. Bellot-Gurlet and P. Dillmann: *J. Raman Spectrosc.*, 2007, **38**, 389–397.
22. C. Rémazeilles and Ph. Refait: *Corros. Sci.*, 2008, **50**, 856–864.
23. F. Gilbert, P. Refait, F. Lévêque, C. Remazeilles and E. Conforto: *J. Phys. Chem. Solids*, 2008, **69**, 2124–2130.
24. S. G. Rees-Jones: *Stud. Conserv.*, 1972, **17**, 39–43.
25. S. Réguer, F. Mirambet, E. Dooryhee, J.-L. Hodeau, P. Dillmann and P. Lagarde: *Corros. Sci.*, 2009, **51**, 2795–2802.

Corroding glass, corroding metals: survey of joint metal/glass corrosion products on historic objects

G. Eggert*

Corroding glass forms alkali(ne) surface films, which may lead to special metal corrosion products in the contact zone, for example different sodium copper carbonates or basic sodium lead carbonate. Sodium copper formate acetate was found on objects exposed to long term emissions from wood. A higher pH value may also lead to basic compounds, which normally do not form, e.g. the newly characterised $Cu_2(OH)_3HCOO$. The alkaline films also create a reactive environment for the formation of formates from formaldehyde (e.g. emitted from glues) due to the Cannizzaro reaction or possibly from the neglected pollutant carbon monoxide. Further products containing, for example, potassium or sulphate may be expected.

Keywords: Glass corrosion, Carbon monoxide, Copper alloys, Copper formates, Sodium copper carbonates, Socoformacite

This paper is part of a special issue on corrosion of archaeological and heritage artefacts

Introduction

Corrosion products on ancient objects and, especially, on copper alloys have been analysed since the rise of modern analytical sciences, one of the first studies being Sage's[1] identification of cuprite on a gilded copper leg found in the Saone in 1766. Despite more than two centuries of research, there are still hitherto unknown corrosion products to be detected. Surprisingly, the eye catching occasional occurrence of green to blue efflorescences at the contact zone between corroding glass and copper alloys (and nowhere else on some objects) was nearly totally ignored until now. To inspire further research, this paper collects the few mentions in the literature together with the author's own unpublished results to picture current knowledge.

Glass corrosion

When on objects made from glass and copper alloys copper corrosion occurs in contact with glass, the latter often shows visible signs of corrosion: hazing, wet surfaces or efflorescences, cracking, roughening, pitting and the like. This gives an obvious hint to the special situation. In glass corrosion, alkali ions bonded in the glass network are leached out by humidity from the air

$$\equiv SiO^- Na^+ + H_2O \rightarrow \equiv SiOH + Na^+ + OH^- \quad (1)$$

The alkaline surface films on corroding glass are maintaining equilibrium with the carbon dioxide content in the air

$$2Na^+ + 2OH^- + CO_2 \rightarrow 2Na^+ + 2CO_3^{2-} + H_2O \quad (2)$$

The reaction can proceed to form some hydrogen carbonate ions as well; the pH value of an equimolar buffer is ~10·1. The special situation of joint glass/metal corrosion is, therefore, simply metal corrosion occurring in the presence of alkali(ne) carbonate containing solutions at ambient atmosphere.

Copper alloys

From the number of observed cases, copper corrosion products outnumber all other metals by far. These are discussed here first; other metals are considered at the end.

Sodium copper carbonates

The first joint copper/glass corrosion product was apparently found by Magee[2] in 1999 on a late fifteenth century enamelled covered copper alloy cup (*tazza*) under a flaking enamel. The enamel consisted of an unstable high alkali/low lime (3·4 mass% CaO) glass composition. Chalconatronite (sodium dicarbonatocuprate(II) trihydrate, $Na_2[Cu(CO_3)_2].3H_2O$) and malachite were identified both in the raised interior and on the exterior of the lid.

Chalconatronite is quite soluble and indicative of copper corrosion in the presence of carbonate/hydrogen carbonate ions. It has first been reported and named by Gettens[3] in 1955, who found it as corrosion product on ancient Egyptian bronzes from arid soda rich soils. It can also be formed by long time immersion of bronzes in sesquicarbonate solution as treatment of 'bronze disease'.[4] Sodium copper carbonate can indeed be synthesised by pouring a copper solution into a sesquicarbonate solution (i.e. a carbonate/hydrogen carbonate buffer). Drying of objects immersed in sodium

Objects Conservation, State Academy of Art and Design, Am Weissenhof 1, D-70191 Stuttgart, Germany

*Corresponding author, email gerhard.eggert@abk-stuttgart.de

Published by Maney on behalf of the Institute
Received 14 December 2009; accepted 20 May 2010
DOI 10.1179/147842210X12754747500603

hydroxide (as in traditional electrolytic treatments)[5] might be another route for its formation.

In contrast to these cases, the *tazza* was never soil logged or immersion treated. Therefore, the sodium in the chalconatronite undoubtedly originates from the glass, making it a real joint corrosion product. In the meantime, another case of chalconatronite has been reported[6] together with cuprite (Cu_2O) on a copper wire inside a fractured mixed alkali lead glass figure made perhaps in Venice in the eighteenth century.

The identification of accretions of another sodium copper carbonate by X-ray diffraction (XRD) (Gandolfi camera), sodium tricarbonatohydroxodicuprate(II) tetrahydrate $Na_3[Cu_2(CO_3)_3(OH)].4H_2O$, is only mentioned in the book on daguerreotypes by Barger and White[7] but was not included in their original research paper.[8] Daguerreotypes (copper plates with a halide sensitised silver plating) are usually sandwiched airtight with a stamped brass mat and a cover glass, which often shows severe signs of corrosion on the inside. Sodium formate (see below) and sulphate have been identified as other crystalline glass corrosion products.

This basic sodium copper carbonate can be synthesised by dropping a copper acetate solution into a sodium hydrogen carbonate solution.[9] Interestingly, this compound containing a hydroxide anion forms at a lower pH value than sodium copper carbonate, which crystallises when a copper solution is added to a carbonate/hydrogen carbonate buffer. In glass corrosion in contact with copper alloys, the pH value of the humidity film on the surface will determine which of the two compounds is formed on drying. This should most likely depend on the glass corrosion rate (i.e. the production of NaOH and its consumption in splitting Si–O–Si bonds) and the deposition rate of acid gases from the environment.

Copper formates

Recently, two cases of sodium copper formate acetate (abbreviated as socoformacite) have been reported by Eggert and co-workers.[10] A green corrosion product was detected during conservation on a seventeenth century limewood box of board games made in the 1670s for Countess Hedwig-Sophie of Hesse-Kassel. Some of the turquoise enamel (made from saline plant ash, about 13–16 mass% Na_2O, wavelength dispersive X-ray microprobe) used for the decoration of the silver alloy (~5·5 mass% Cu, atomic absorption spectroscopy) backgammon fields flaked off. In some of the now open silver grooves, where there was formerly enamel (and only there!), socoformacite was identified by XRD. The same holds true for an about 100 year old Chinese theatre hat (Ethnological Museum of Heidelberg, inventory no. 30406), which has long been stored in a wooden cupboard made from oak. The hat consists mainly of silk stabilised with cardboard. Silvered copper wire and glass beads were used as decorative elements. One metal spiral was covered with a light green blue corrosion product where it was in direct contact with an apparently weathered glass bead, showing iridescence and crizzling. As wood is known to continuously emit traces of formic and acetic acid (ester bound in its hemicelluloses) into the air, this might reasonably be the source for formate and acetate in joint copper/glass corrosion products.

The identification by XRD was at first impossible, as there is no entry for it in the powder diffraction file, and the crystal structure is unknown. The match of the author's diffraction pattern with diffraction data published by Scott[11] finally tracked to the study of this compound by Trentelman *et al.*[12] (see the comparative table of data in Eggert *et al.*).[10] According to their X-ray photoelectron spectroscopy and Raman spectroscopy data, socoformacite is a mixed sodium copper (1 : 1) formate acetate, with the formate/acetate ratio estimated between 1 : 2 and 2 : 1. The general formula might, therefore, be given as $NaCu(HCOO)_{1+X}(CH_3COO)_{2-X}$, with $0 \leqslant X \leqslant 1$. Trentelman *et al.* observed it on a number of archaeological copper alloy artefacts containing no glass from various cultures and a number of collections. Apparently, all these finds were exposed to carbonyl pollutants in the museum atmosphere as is often the case when, for example, wood is used for building or display cases. Only one object was non-archaeological, a fifteenth century Italian leaded tin bronze statue.[12] It is not clear if socoformacite was formed by the action of acid gases on already existing chalconatronite (from soil logging or conservation treatment) or simply crystallised from solutions containing all four ions formed by corrosion in humid air. Eggert *et al.* were unable to prove either route in model experiments or even to grow single crystals.[10]

As Trentelman *et al.*[12] proved that socoformacite was not extremely rare on museum artefacts without glass, it was an interesting question to find out how rare these newly reported cases with sodium from glass corrosion would be. A call for samples in conservation conferences and newsletters lead to the identification of socoformacite in seven further combined metal/glass objects of totally different backgrounds within 3 years:[13]

(i) a traditional Black Forest *Schäppel* (girl's headdress worn on high ecclesiastical holidays and wedding) decorated with glass beads from Lauscha (many mirrored) on wires showed copper contact corrosion. *Schäppels* were of course stored at home in wooden cupboards or boxes

(ii) a Christmas glass ball (also possibly made in Lauscha, in the collection of Badisches Landesmuseum Karlsruhe), silver mirrored inside, was decorated on the outside with silvered

Table 1 Joint soda glass/copper alloy corrosion products

Chemical name	Formula	Other names	References
Sodium dicarbonatocuprate(II) trihydrate	$Na_2[Cu(CO_3)_2].3H_2O$	Chalconatronite, sodium copper carbonate	2,6
Sodium tricarbonatohydroxodicuprate(II) tetrahydrate	$Na_3[Cu_2(CO_3)_3(OH)].4H_2O$	Basic sodium copper carbonate	7
Sodium copper formate acetate	$NaCu(HCOO)_{1+X}(CH_3COO)_{2-X}$	Socoformacite	10,13
Dicopper trihydroxy formate	$Cu_2(OH)_3HCOO$	Basic copper formate	14

copper wire. Green corrosion products (not analysed) had been stripped with a chelating agent (Na_2EDTA) in 2002. When inspected again 7 years later, the corrosion products reoccurred where the wire was in contact with the glass and were identified as socoformacite
(iii) two enamel objects made in Limoges from the Museum für Angewandte Kunst Frankfurt had socoformacite and basic copper formate (see below) efflorescences at the border between the metal and enamel. Acetate containing efflorescences ('thecotrichite') occurred also on ceramics from this collection
(iv) the brass passepartout of a glass framed daguerreotype from the 1860s (donated to the author by Pau Maynes) showed green corrosion spots. If stored flat, condensing humidity might have dropped down from the glass
(v) a small female enamel figure from a 'Handstein' (the second half of the sixteenth century, Kunsthistorisches Museum Vienna) with interior copper wire showed socoformacite efflorescences growing from fractures in the arm
(vi) a glass cabochon (imitation of gem) with a gilded silver mounting (the thirteenth century, apparently debased with copper) on the cover of the Otto-Adelheid-Evangeliar (Quedlinburg) developed socoformacite in the contact zone. It was formerly displayed in a case with high emissions of acetic and formic acid.

Undoubtedly, publication of these occurrences will lead to further identifications on combined glass/metal objects.

In the meantime, sampling in search of socoformacite has lead to the identification of another corrosion product,[14] dicopper trihydroxy formate $Cu_2(OH)_3HCOO$. The author's diffraction data (not in the PDF) could be matched with a basic copper formate given by Scott *et al.*[15] as their film 770 (Table 3). Single crystals of this compound could be synthesised during copper corrosion experiments with formic acid and ammonia, adjusted to pH 9·1, and the crystal structure could be determined.

The compound was identified on:
(i) a 200 year old glass flute with silver mounting (Rijksmuseum Amsterdam)
(ii) a Baroque silver mounted ruby glass box (Green Vault Dresden)
(iii) the enamelled Limoges objects from Frankfurt (see before).

Corrosion experiments with copper coupons over formic acid only yield neutral formates or perhaps also $Cu(OH)HCOO$.[16,17] The formation of the trihydroxy compound is due to the glass corrosion as it needs a higher pH.

Copper formates (without sodium) on historic objects are very rare; only two other cases have been found in the literature.[18,19] The connection with glass corrosion reported here might be significant, as the alkaline surface film not only absorbs formic acid but may also form it from formaldehyde (emitted, e.g. from glues) without further oxidation step by disproportionation (Cannizzaro reaction)[20]

$$2H_2CO + NaOH \rightarrow CH_3OH + NaHCOO \quad (3)$$

Another possible source for formate is the hitherto neglected air pollutant carbon monoxide. This gas does not react in water and is not the anhydride of HCOOH (as hydrogen is bonded to carbon). However, Berthelot[21] had already discovered in 1856 that alkali hydroxide solutions can absorb carbon monoxide to form formates. The modern synthesis of sodium formate uses heat and pressure according to

$$NaOH + CO \rightarrow NaHCOO \quad (4)$$

Currently, experiments are set up with model glasses rich in sodium exposed to CO to see how relevant this reaction is. Carbon monoxide concentrations in rooms where organic material is burnt (fireplaces, stoves, tobacco smoking, etc.) are in the order of some 10 ppm (!). Historically, this might be much more relevant for formate formation than other trace gases in the parts per billion range.

Other metals

With copper forming up to now four different identified corrosion products (and some still unidentified!) in contact with corroding glass, what about the other ancient metals? Gold and the pre-Columbian platinum objects of course do not corrode at all in the museum. Observed corrosion on silver alloys was always due to the minor copper content. There is no stable silver carbonate, and the formation of Ag_2O needs such a high pH value not likely to occur even in contact with the corroding glass in the presence of carbon dioxide from air. Tin also does not form a stable carbonate but corrodes to (hydr)oxides even at normal pH value, so special products from glass corrosion are unlikely. From the potential–pH stability diagrams, the same holds true for iron in the presence of oxygen, siderite [iron(II) carbonate] is found only under strongly reducing conditions in the soil.[22] Steel tends to passivate at higher pH value and is used for containers of alkali carbonate solutions. The mutual influence of glass and steel corrosion is of special interest for the long term storage of vitrified nuclear waste.[23]

However, the chemistry of lead corrosion products is as rich as that of copper. Basic sodium lead carbonate $NaPb_2(OH)(CO_3)_2$ has recently been reported as white pigment,[24] possibly formed by a non-traditional attempt to synthesise lead white (basic lead carbonate, hydrocerussite) by precipitation of a soluble lead salt (nitrate or acetate) with sodium carbonate. Solutions containing more than ~0·15 mol L^{-1} Na^+ form this compound, not hydrocerussite.[25] Such conditions may occur during

Table 2 Possible ions and their sources in joint glass/copper alloy corrosion products*

Source	Agents/ precursors	Cations	Anions
Copper alloys		Cu^{2+} (Cu^+)	
Glass		Na^+ (K^+) (Ca^{2+}) (Mg^{2+}) (H^+)	OH^- (O^{2-})
Wood, glue, etc.	H_2CO, $HCOOH$ CH_3COOH		$HCOO^-$, CH_3COO^-
Air	O_2, H_2O, CO_2		CO_3^{2-}
	SO_2, NO_x		(NO_3^-)(SO_4^{2-}),
	CO		$HCOO^-$?

*( ) ions so far not observed in joint corrosion products; ? hypothesis needing expert verification.

Table 3 Powder diffraction data for dicopper trihydroxy formate (*I*=relative intensity, ‰)

Calculated from single crystal structure[14]		Sample from baroque ruby glass box (see text)		Scott *et al*. synthetic compound[15]	
d, Å	*I*	*d*, Å	*I*	*d*, Å	*I*
				7·35	30
6·6659	1000	6·6970	1000	6·64	1000
5·3745	7	5·3837	9	5·39	50
4·4824	1	4·5030	4	4·52	50
4·0198	19	4·0316	22	4·08	100
3·8049	14	3·8034	13	3·82	50
3·7154	7	3·7401	10	3·71	30
3·3330	158	3·3485	250	3·37	600
3·2616	7	3·2563	6	3·31	20
3·1669	21	3·1861	15	3·19	50
3·0280	3				
2·9200	10	2·9335	8		
2·8716	4				
2·7698	106	2·7695	18		
2·7569	7	2·7637	136		
2·6872	21	2·6918	30		
2·6381	150	2·6483	199		
2·5745	27	2·5804	30		
2·5377	6	2·5560	12		
2·5188	<1	2·5162	5		
2·4563	2	2·4616	16		
2·4454	122	2·4366	180	2·48	100
2·3472	207	2·3598	245		
2·3405	<1	2·3564	56		
2·2869	4	2·2961	32		
2·2844	23	2·2876	8	2·28	200
2·2675	1	2·2515	3		
2·2412	5	2·2323	26		
2·2220	16	2·2228	71	2·22	100
2·2191	57				
2·1394	<1				
2·1374	4				
2·0860	<1				
2·0437	3				
2·0099	<1				
2·0035	64	1·9979	94	1·99	200
1·9450	124	1·9568	178	1·93	50
1·9320	<1	1·9408	6		
1·9025	<1	1·8977	<1		
1·8898	<1	1·8924	3		
1·8794	<1	1·8722	1		
1·8660	<1	1·8700	30	1·87	50
1·8630	<1	1·8610	5		
1·8577	24	1·8326	5		
1·8249	45	1·8282	61	1·82	100
1·8237	1				
1·7970	1				
1·7949	<1				
1·7914	<1				
1·7806	1	1·7796	4	1·76	50
1·7760	2				
1·7738	<1				
1·7318	<1				
1·7267	<1				
1·7227	<1				
1·7179	1	1·7186	6		
1·7165	<1				
1·6709	<1				
1·6665	13	1·6743	26	1·68	100
1·6651	1	1·6699	2		
1·6314	<1	1·6349	5		
1·6308	19	1·6289	9		
1·6303	1	1·6282	23		
1·6140	1	1·6244	10		
1·6068	<1	1·6197	11		
		1·6142	8		
1·5968	40	1·6068	47	1·61	100
		1·5931	7		
1·5867	61	1·5875	99		
1·5834	<1				

Table 3 Continued

Calculated from single crystal structure[14]		Sample from baroque ruby glass box (see text)		Scott *et al.* synthetic compound[15]	
d, Å	*I*	*d*, Å	*I*	*d*, Å	*I*
1·5798	<1				
1·5747	<1				
1·5742	<1				
1·5669	<1				
1·5568	<1				
1·5453	29	1·5456	36	1·55	50
1·5418	38	1·5437	64		
1·5198	15	1·5307	17		
1·5140	24	1·5209	9		
1·5134	<1	1·5199	28		
1·5033	35	1·5066	46		
1·4941	<1	1·5010	6		
1·4809	<1	1·4907	3		
1·4764	34	1·4831	48	1·48	30

soda glass corrosion in contact with lead (e.g. leaded windows). In tiny gaps between lead and glass, corrosion solutions are not washed away and could easily develop higher sodium concentrations. Indeed, lead foil (Merck, Darmstadt, Germany; p.a., 1·07365·000, Cu<0·002%) immersed into a 1 mol L^{-1} solution of sodium carbonate (Merck, p.a., 1·06392·0500) in the presence of air developed a brownish green (!) corrosion. A sample analysed by XRD after 3 days contained basic sodium lead carbonate as the only crystalline corrosion product as expected; a spot test for copper was negative. Currently, a corrosion experiment is performed with lead foil in contact with soda glass over half a year.

Corroded old leadings of historical soda glass windows need to be systematically analysed to prove whether basic sodium lead carbonate is formed on them under natural weathering conditions.

Conclusions

Since the start of this research on joint metal/glass corrosion products in 2006, already four copper (Table 1) and possibly one lead compound have been identified. Glass corrosion leads to an alkali(ne) environment which provides (Table 2) the following aspects.

1. Alkali cations, which may be precipitated in mixed corrosion products. So far, only sodium compounds have been identified, but there is no reason why potash glasses, which tend to be even more unstable than soda glasses, could not react similarly. Earth alkaline ions like Ca^{2+} are known to occur in glass corrosion products and, therefore, might also occur in joint products.
2. An alkaline pH, which could result in basic compounds not observed under more neutral conditions.
3. Alkaline surface films, which readily absorb acid gases like carbon dioxide, formic and/or acetic acid and may result in the precipitation of carbonates, formates and/or acetates on drying. Although sulphates form during glass corrosion, e.g. on church windows [gypsum, $Ca_2SO_4.2H_2O$, and syngenite, $K_2Ca(SO_4)_2.H_2O$], due to air pollution with SO_2, no mixed glass/metal sulphates have been detected so far. On objects where corrosion products are not washed away by rain or cleaning, even nitrates (from NO_x in the air) might be a possibility.
4. A reactive environment for the formation of formates from formaldehyde due to the Cannizzaro reaction or possibly from carbon monoxide.

As, so far, only few of the possible combinations of ions (Table 2) have been found in joint corrosion products and some compounds could not be identified yet, sampling of corrosion products on historic objects in the contact zone between metal and glass will be continued. Raman microscopy will aid in the identification of ions and compounds. Corrosion products contain some otherwise unavailable information to what conditions artefacts were exposed over centuries.

Acknowledgements

The author is grateful to Bruno Barbier and Harald Euler for all XRD measurements. This study would not be possible without many conservation colleagues and students keeping an open eye for the author for interesting samples.

References

1. M. Sage: *Obs. Phys. Hist. Nat. Arts*, 1779, **14**, 155–157.
2. C. E. Magee: Proc. ICOM-CC 12th Triennial Meet., (ed. J. Bridgland), Vol. 2, 787–792; 1999, London, James & James.
3. R. J. Gettens: *Stud. Conserv.*, 1955, **2**, 64–75.
4. A. M. Pollard, R. G. Thomas and P. A. Williams: *Stud. Conserv.*, 1990, **35**, 148–152.
5. H. Plenderleith and A. E. A. Werner: 'The conservation of antiquities and works of art', 2nd edn; 1971, London, Oxford University Press.
6. R. J. G. Sobott: Proc. 1st Workshop on 'Archäometrische 3D-Marker archäologischer Gläser und Keramiken', Leipzig, Germany, October 2007, University of Leipzig.
7. M. S. Barger and W. B. White: 'The Daguerreotype', 167; 1991, Washington, DC, Smithsonian Institution Press.
8. M. S. Barger, D. K. Smith and W. B. White: *J. Mater. Sci.*, 1989, **24**, 1343–1356.
9. A. K. Sengupta and A. K. Nandi: *J. Inorg. Nucl. Chem.*, 1974, **36**, 2479–2484.
10. G. Eggert, A. Wollmann, B. Schwahn, E. Hustedt-Martens, B. Barbier and H. Euler: Proc. ICOM-CC 15th Triennial Conf., (ed. J. Bridgland), Vol. I, 211–216; 2008, New Delhi, Allied Publishers.
11. D. A. Scott: 'Copper and bronze in art: corrosion, colorants, conservation', 301, 446; 2002, Los Angeles, CA, Getty Conservation Institute.
12. K. Trentelman, L. Stodulski, D. Scott, M. Back, S. Stock, D. Strahan, A. R. Drews, A. O'Neill, W. H. Weber, A. Chen and S. Garrett: *Stud. Conserv.*, 2002, **47**, 217–227.
13. G. Eggert, A. Bührer, H. Euler and B. Barbier: in 'Glass and ceramics conservation 2010', (ed. H. Römich); 2010, Corning, NY, Corning Museum of Glass. In press.
14. H. Euler, B. Barbier, A. Kirfel, S. Haseloff and G. Eggert: *Z. Kristallogr. NCS*, 2009, **224**, 609–610.
15. D. A. Scott, Y. Taniguchi and E. Koseto: *Rev. Conserv.*, 2001, **2**, 73–91.

16. H. Gil and C. Leygraf: *J. Electrochem. Soc.*, 2007, **154**, C611–C617
17. D. M. Bastidas, V. M. La Iglesia, E. Cano, S. Fajardo and J. M. Bastidas: *J. Electrochem. Soc.*, 2008, **155**, C578–C582.
18. L. Robinet and D. Thickett: in 'Raman spectroscopy in archaeology and art history', (ed. H. G. M. Edwards and J. M. Chalmers), 325–334; 2005, Cambridge, RSC Publishing.
19. U. Kugler: *Met Object.*, 2003, **5**, 6–8.
20. L. Robinet, C. Hall, K. Eremin, S. Fearn and J. Tate: *J. Non-Cryst. Solids*, 2009, **355**, 1478–1488.
21. M. Berthelot: *Ann. Chim. Phys.*, 1856, **46**, (3), 477–491.
22. D. A. Scott and G. Eggert: 'Iron and steel in art: corrosion, colorants, conservation', 53–56; 2009, London, Archetype Publications.
23. G. de Combarieu, P. Barboux and Y. Minet: *Phys. Chem. Earth*, **32**, (1–7), 346–358.
24. H. Kutzke, S. Heym and A. Schönemann: Proc. Annual Meet. on 'Archäologie und Denkmalpflege 2009', Munich, Germany, March 2009, Metalla (Sonderheft), Vol. 2, 252–253.
25. F. Auerbach: *Z. f. Elektroch.*, 1913, **19**, 827–830.

Long term assessment of atmospheric decay of stained glass windows

T. Lombardo*[1], C. Loisel[2], L. Gentaz[1], A. Chabas[1], M. Verita[3] and I. Pallot-Frossard[2]

Several studies indicate that the decay of medieval stained glass windows is related to both the glass composition and the characteristics of the environment. The kinetics of the decay processes has been always described through experiences performed in aqueous confined conditions, which are obviously not encountered in real condition. A research programme has been set up in order to assess the kinetics of atmospheric weathering. The long and short term weathering has been studied through respectively ancient glass fragments and glass analogues exposed in the field. This paper presents the preliminary results of the long term. A characterisation of the morphology and chemical modification induced by weathering has been performed. Results will be used in the next step to build up a model of long term atmospheric weathering kinetics.

Keywords: Stained glass windows, Atmospheric environment, Weathering

This paper is part of a special issue on corrosion of archaeological and heritage artefacts

Introduction

As any other kind of material glasses constituting stained glass windows undergo a more or less severe damage, which results from the interaction between the glass matrix (intrinsic factors) and the environment (extrinsic factors).

Several researches have been performed to understand the weathering of silicate glasses in aqueous solutions, especially in confined conditions[1–14] demonstrating that glasses undergo two kinds of deterioration processes: leaching (selective dissolution) and corrosion (congruent dissolution). In the case of leaching, the cations (glass modifiers) present in the glass matrix are gradually extracted from the network and substituted by hydrogenated species (H^+, H_3O^+, H_2O) according to a diffusion process[1,2] and obeying to a square root kinetics.[1,3–5] Leaching leads to the formation of a hydrated layer, enriched in silanol groups (≡Si–OH) and depleted in cations, at the upper most surface of the glass.[1,6]

The corrosion process takes place in basic condition, leading to the destruction of the siloxane groups (Si–O–Si) forming the glass network[3–5,7] and to the precipitation of corrosion products (formation of a gel layer in which ion diffusion is quite active). The kinetics of corrosion in confined aqueous solution is linear.[5] While corrosion acts on all glasses no matter their composition,[8] leaching is strongly dependent on it. In general, pure silica glasses are considered inert; the progressive addition of cations results in durable (low proportion of cations) and low durable ones (higher proportion of cations).[7,9–11] These two processes are strongly influenced by the pH of the solution.[3–5,7,8,10,12,13]

Glasses used in the medieval age to build up stained glass windows, were usually made of silica and high amount of alkali and alkaline-earth elements (mainly K, Ca, Mg, Na). The result is that most of these glasses, considered as low durable ones, underwent weathering processes because of their interaction with the atmosphere. Evidence of both leaching and corrosion are largely found in ancient medieval stained glass windows.[14–23] Studies have shown that a hydrated leached layer is formed and progressively a fissure network is formed into it, allowing the water to diffuse and the weathering to progress toward the bulk glass in a finger-like manner.[8,18] In general, microcracks are area favourable to the development of weathering.[8] Corrosion attack results in a more or less uniform pitting of the glass surface. Pits can also form inside the uppermost modified layer. Beside leaching and corrosion, new deterioration forms were observed on ancient stained glass windows; indeed gypsum crusts developed at their outdoor surface.[15–19] Crusts are formed by a complex interaction between the cations extracted from the glass, atmospheric gases (CO_2, SO_2)[18,20–22] and particulate matter.[17] Finally, several researches have demonstrated that atmospheric pollutants, climate parameters (temperature, relative humidity and rain), are involved in all weathering processes affecting medieval glasses.[20–25]

Although a lot of researches have been performed in the domain of the deterioration of stained glass window glasses, several issues need to be deeply investigated. In particular, the kinetics of the glass atmospheric weathering has to be determined, as up to now all the studies are based on the results obtained in aqueous solutions.

[1]LISA, Universités Paris 12 et Paris 7, CNRS, France
[2]Laboratoire de Recherche des Monuments Historiques, LRMH, France
[3]Laboratorio di Analisi dei Materiali Antichi, LAMA, Italy

*Corresponding author, email tiziana.lombardo@lisa.univ-paris12.fr

Received 15 December 2009; accepted 5 April 2010
DOI 10.1179/147842210X12710800383800

To answer this question, a program funded by the French Ministry of Culture and Communication (PNRCC) has been set up. It is based on a field exposure of analogues glasses joined with an environmental survey of several key parameters to access the short term deterioration; while long terms deterioration has been studied on ancient medieval glasses issued from different sites. This paper focuses on the results of the long term weathering and illustrates several examples of ancient deteriorated glass fragments issued from fourteenth century stained glass windows.

Materials and methods

Fragments of glass of stained glass windows come from two French historical monuments: the cathedral of Notre-Dame of Evreux (three samples, Ev1–3) and the Abbey of Saint-Ouen in Rouen (five samples, Ou1–5) (Table 1). In both cases, samples are dated back to the fourteenth century and present evidence of weathering, more or less intense. They correspond to glasses removed during restoration work in the 60–70 s, and stored in appropriate conditions, at the Laboratoire de Recherche des Monuments Historiques.

Only uncoloured glasses have been selected as they were more abundant and because of their similarity with the uncoloured analogue glasses exposed in the field to access weathering in the short term. A first observation of these fragments performed under optical microscopy (Leica DMRM) allowed eliminating Ou1 because it is characterised by numerous scratches indicating that a cleaning by brush was performed during previous restoration at an undefined date. As the main goal of this research is to assess the kinetics of weathering, an exact estimation of the time span during which the weathering took place is essential. Of course, for all the other samples, an error on the weathering timing is possible, because of previous restorations, reuse of glass pieces, flaking off of microfragments. Therefore, results have to be interpreted carefully.

On each of the four samples, thus selected (Ev1, Ou2, Ou4, Ou5), two small fragments (a, b), (about 1–1·5 cm^2 area) have been sampled, far enough from the lead frame, and prepared as cross-sections to be further investigate by scanning electron microscopy (SEM). Analyses have been performed using a JEOL JSM-6301 F linked with EDX (Link ISIS 300, Oxford) allowing the detection of elements with $Z \geqslant 5$. The analytical conditions were as follows: acceleration tension: 20 kV; working distance: 15 mm; dead time: 20–30%; energy calibration: K_{α} Cu.

The theoretical durability of each glass has been estimated through the calculation of both NBO/T (Non-Bridging Oxygen/Tetrahedrons)[26] and ΔG_{hyd} (free hydration energies)[26–28] values using the SEM–EDX chemical composition of the bulk glass.

For the exact quantification of the thickness of the modified layer, images of the entire sample, taken by SEM in backscattering mode at a constant magnification ($\times 200$ to $\times 400$), were processed using image analysis software (Histolab, Microvision). For each photo, the area occupied by the modified layer (darker than the bulk glass) was measured, and then the average thickness was calculated based on the approximation of a rectangular homogeneous layer. The operation was repeated for each photo. Although this calculation smoothes part of the heterogeneity of the modified layer (in particular, it might neglect the rare areas with no visible weathering), it has the advantage of a more rapid and complete investigation of the whole sample.

Results and discussion

Elemental analyses (SEM–EDX) on cross-sections of samples Ou2b, Ou4b, Ou5a and Ev1b, allowed accessing the chemical composition of the bulk glass (Table 2).

Whatever the glass, their composition is quite complex: Si, K and Ca are the main constituents. Furthermore, they contain a large spectrum of elements with similar concentrations for all glasses: magnesium (~7%), phosphorus (3·6–4·5%); sodium (about 1·9–2·5%), aluminium and manganese (0·7–1·3%). Iron was detected in low amount in all samples (0·3–0·8%) with the exception of Ou4b. Traces of chlorine are observed in Ou5a (~0·4%) (not reported in Table 2) and lead (0·1%) is found in Ou2b. Although, sulphur has been detected, the high standard deviation values do not allow any conclusion.

Table 1 Macroscopic description of fourteenth century glass fragments

Provenance	Name	Colour	Dimension (shape)	Observations
Notre Dame d'Evreux Cathedral	Ev1	Uncoloured	12 × 6 cm (pentagon)	Opaque external face. Pits on the internal face only
	Ev2	Red	9 × 3·5 cm (triangular)	Opaque external face and several scratches at the internal face
	Ev3	Uncoloured + silver stain	7 × 3 cm (rectangular)	Opaque external face, corrosion pits at both internal and external faces
Saint Ouen Abbey in Rouen	Ou1	Uncoloured	18 × 7 cm (rectangular)	Internal face more opaque than external one. Pits, craters and irisation at the external face, regular and numerous scratches are also observed.
	Ou2	Uncoloured	15 × 6 cm (lozenge)	Both external and internal faces are opaque and present craters. A grisaille is present at both faces.
	Ou3	Yellow	6 × 4·5 cm (pentagon)	Numerous craters on the external face, rare craters on the internal one
	Ou4	Uncoloured	7 × 3 cm (irregular)	Opaque external face. Pits on the internal face
	Ou5	Uncoloured	7 × 4 cm (rectangular)	Opaque external face. Pits on the internal face only

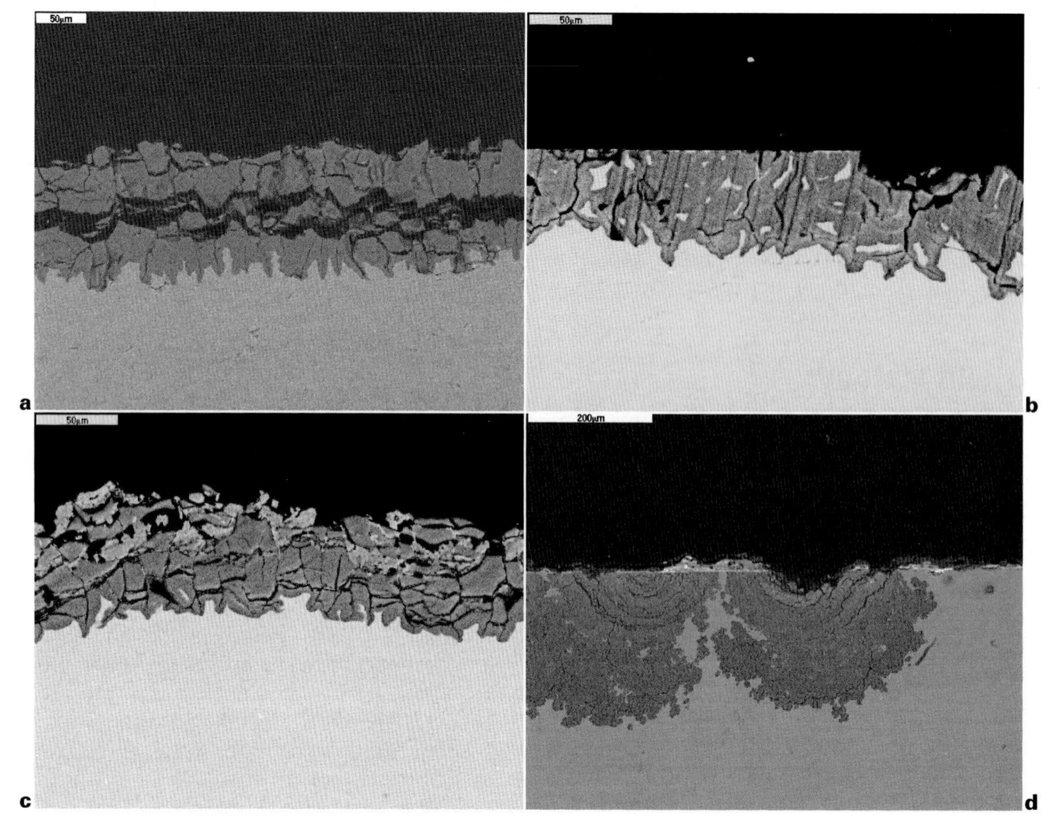

1 Images (SEM) (back scattering mode) of modified layers present at external surface of samples *a* Ev1b, *b* Ou4b, *c* Ou5a and *d* Ou2b

In general, all four glasses are characterised by low silicon content (<57·8%) and quite high proportion of calcium and potassium, indicating their low durability, as confirmed by their high NBO/T (higher than 1) and ΔG_{hyd} (between −13·2 and −16·6 kcal mol^{-1}) values (Table 3). For comparison, for a durable Si–Ca–Na (Planilux) NBO/T and ΔG_{hyd} are 0·82 and −4·4 kcal - mol^{-1} respectively.

All samples present a modified layer at the uppermost surface at their external and internal sides. However, only results on external side, in contact with the outdoor atmospheric environment, are presented. The investigation of the morphology of these layers shows a quite complex situation; although some similarities exist, a certain internal (for the same sample) and external (among different samples) heterogeneity is observed. Indeed, all samples present a more or less continuous layer (Fig. 1*a–d*) characterised by an extremely variable thickness (Table 3) as deeply deteriorated zones coexist with less weathered or rare non weathered ones. In backscattering mode, for all samples, these layers are darker when compared to the bulk glass (Fig. 1), and are

Table 2 Elemental composition (mean and standard deviation, σ) of bulk glass, of leached layer (LL) and Δbulk=([LL]−[Bulk])/[Bulk]) (in wt-% of oxides) of samples Ou2b, Ou4b, Ou5a and Ev1b analysed by SEM–EDX

			Na_2O	MgO	Al_2O_3	SiO_2	P_2O_5	SO_3	K_2O	CaO	MnO	Fe_2O_3	PbO
Ou2b	Bulk	Mean	2·53	6·81	0·98	57·76	3·6	0·18	12·7	13·79	0·70	0·6	0·1
		σ	0·17	0·48	0·18	2·64	0·26	0·15	1·23	0·92	0·12	0·15	0·01
	LL	Mean	1·01	1·14	1·65	74·09	2·87	0·86	1·38	11·43	2·28	1·00	2·35
		σ	0·24	0·74	0·27	7·35	1·55	0·45	1·1	5·15	2·26	0·25	2·09
	Δbulk		−0·6	−0·83	0·69	0·28	−0·2	3·82	−0·89	−0·17	2·25	0·66	23·7
Ou4b	Bulk	Mean	1·89	6·80	1·33	54·64	4·52	0·26	17·05	12·34	1·09	...	...
		σ	0·26	0·59	0·09	2·41	0·22	0·18	1·03	0·63	0·13	...	...
	LL	Mean	0·70	1·87	2·01	82·29	3·91	2·47	3·59	2·21	0·98	...	1·18
		σ	0·25	0·57	0·12	4·98	1·57	0·67	2·2	2·04	0·42	...	1·18
	Δbulk		−0·63	−0·73	0·51	0·51	−0·14	8·45	−0·79	−0·82	−0·1	...	...
Ou5a	Bulk	Mean	2·59	7·03	1·30	53·61	4·5	0·28	16·73	11·73	1·02	0·76	...
		σ	0·23	0·33	0·17	2·86	0·19	0·23	1·80	1·01	0·11	0·11	...
	LL	Mean	1·04	1·82	1·96	77·89	4·16	5·11	1·80	4·13	0·81	1·12	...
		σ	0·29	0·94	0·34	15·3	1·3	10·1	1·78	6·65	0·47	0·24	...
	Δbulk		−0·6	−0·74	0·5	0·45	−0·1	17·2	−0·58	−0·89	−0·65	−0·21	...
Ev1b	Bulk	Mean	2·49	7·33	0·73	52·98	4·15	0·20	16·93	13·4	1·33	0·31	...
		σ	0·09	0·11	0·11	0·512	0·18	0·06	0·52	0·18	0·11	0·16	...
	LL	Mean	1·14	2·70	1·09	82·46	1·97	2·19	4·48	3·02	0·64	0·58	...
		σ	0·68	2·07	0·27	14·6	1·14	0·98	6·3	5·38	0·48	0·26	...
	Δbulk		−0·54	−0·63	0·49	0·56	−0·53	9·89	−0·74	−0·77	−0·52	0·86	...

very likely enriched in hydrogenated species, suggesting a leaching process. The weathering front (boundary between the modified layer and the bulk glass) present a finger like morphology, associated with a network of fissures perpendicular to the glass surface and acting as preferential paths through which atmospheric fluids can diffuse inside the bulk glass and eventually react with glass cations to form precipitates. For three samples (Ou4b, Ou5a and Ev1) the layer is characterised by an intense fissure network (parallel and perpendicular to the original surface) (Fig. 1*a–c*) isolating portion of the leached layer in a puzzle like pattern. Sulphur rich deposit appears at the uppermost part of sample (Ou4b and Ev1b) or inside the fissure network (Ou5a, Fig. 1*c*). Moreover, the original surfaces of these samples is, in some cases, missing testifying of a flaking off of a brittle leached layer leading to an extreme variability of the thickness and at its quite limited extent. Sample Ou2 differs significantly from the other, indeed its layer has a laminated pattern and each lamina (few nm thick) seems to have a different chemical composition (Ca enrichment). Some pits and craters, at different stages of their formation, are also observed inside the leached layer (Fig. 1*d*). They are formed in the later stages of weathering and seem more related to corrosion processes. With the exception of pitted areas, the original surface of this glass is quite well conserved, probably protected by the presence of an exogenous deposit or of a grisaille. As a consequence, the thickness of the modified layer measured on this sample is much higher than for the other ones (maximum 228 μm), despite the fact that, its chemical durability, in view of its NBO/T, ΔG_{hyd}, SiO_2, K_2O and CaO concentrations, is higher than for the other three glasses. Indeed, for these latter, the lower thickness values only give a partial view of the weathering experienced as an important loss of material is observed. Several hypotheses could be addressed to explain this loss. First, it is due to a mechanical process occurred during either the sample preparation or a past restoration (not documented). Second, the flaking off of these leached layers might testify of the fact that, since the beginning of its formation, they have a weak structure (e.g. more porous), thus during their growth, fractures can easily develops isolating portions of it and causing an intense flaking off. These hypotheses are in agreement with the higher theoretical low durability (based on their chemical composition) of these three samples.

The chemical characterisation of these layers has been undertaken, the mean values and their standard deviations are reported in Table 2. It has to be noticed that, due to the technique used, these values have to be considered as indicative of the chemical changes occurred on glass and not as a quantification of the exact composition of the leached layers. For instance hydrogenated species cannot be quantified by SEM–EDX.

An overall picture of the situation can be drawn. Indeed, for all samples, the mean composition of these layers compared to the bulk glass is strongly depleted in K, Ca, Mg, Na, P and Mn (except Ou2b), and relatively enriched in Si and slightly in Al. They are also generally enriched in Fe except for Ou5a. A general S enrichment (atmospheric supply) is observed on all samples, associated or not with crystal precipitations. A gradient of concentrations is detected inside these layers (high standard deviation values), the modification (depletion–enrichment) being, for most of the elements, the strongest close to the glass surface and attenuating towards the bulk. Frequently, Ca behaviour differs since its concentration strongly fluctuates inside the leached layer alternating Ca highly depleted zones to areas of intense Ca accumulation (especially for Ou2b). In some cases, these latter corresponds to areas of S enrichment with crystal aggregates (gypsum or anhydrite?) precipitated in the fissure network present in the leached layer (Fig. 1*c*). In other cases, no visible crystallisations are observed in the Ca enriched areas, which appear as laminas or isolated microspots brighter (in backscattering mode) than the surrounding layer (Fig. 1*a*), the enrichment in silicon is relatively less important in these areas. Finally, a lead enrichment is observed on sample OU2b and very luckily linked to the diffusion of this element from the grisaille present at the glass surface.

Comparing the mean concentrations of the leached layer and of the bulk, K extraction varies from 70 to 90% (slightly less for Ou5a). This similarity, allow rejecting the hypothesis of a removal of the leached layer during the cross-section preparation, as, due to the observed chemical gradient, lower values should be found in the case of removal of the uppermost part of the sample Ou4b–5a and Ev1b. Na and Mg concentrations are reduced of 50–80% depending on the glass. Phosphorus concentrations decrease of ~15% for Ou2–4–5 and ~52% for Ev1. Calcium modifications vary from highly depleted samples (Ou4 and Ev1, 80–90%) to less depleted one (Ou2b, 17%). This latter observation is in agreement with the fact that an important variation of Ca is observed in the modified layer and suggests of a re-precipitation of this element or a differential extraction.

Table 3 NBO/T, ΔG_{hyd} (kcal mol^{-1}) and thickness of leached layer (mean, min., max., standard deviation σ and median) of samples OU2b, OU4b, Ou5a and Ev1b

			Thickness of the leached layer, μm			
Sample	NBO/T	ΔG_{hydr}	Mean	Min.	Max.	σ
Ou2b	1·08	−13·2	83·2	5·8	228·9	51
Ou4b	1·14	−16·1	36·3	18·8	52·9	10·2
Ou5a	1·17	−15·6	40·9	0·0	79·6	22·9
Ev1b	1·32	−16·6	66·6	41·8	83·2	15·4

Conclusions

In order to study the weathering processes affecting medieval glass induced by their interaction with the surrounding atmosphere, fragment of ancient stained glass windows (fourteenth century) issued from two historical buildings, the Cathedral of Notre-Dame of Evreux and the Saint-Ouen Abbey in Rouen, have been analysed by SEM–EDX.

All glasses are Si–Ca–K in composition and contain quite constant proportions of other elements (Mg, Na, P, Mn, Al, Fe). They are all characterised by the presence of a leached layer depleted in cations such as K, Ca, Na, Mg, P and Mn, and relatively enriched in Si and Al and Fe. The morphology of this layer is different for each glass and presents a more or less homogenous distribution within the sample. Extremely fractured and brittle layers have been observed as well as compact and laminated ones. Evidence of corrosion has been found in this latter.

At a first glance, these preliminary results seem to indicate a discrepancy between the theoretical durability of the glass and the observed weathering, showing that the situation is quite complex as a large number of factors (different initial glass structures, variety of microenvironments, undocumented restorations, etc.) are involved in weathering processes. Therefore, a large panel of sample is necessary to better quantify long term glass weathering. All results, thus obtained, will be integrated in a model describing atmospheric weathering kinetics.

Acknowledgement

The authors thank the French Ministry of Culture and Communication for funding this research via the 'national research program for the comprehension and conservation of Cultural Heritage materials' (PNRCC).

References

1. R. H. Doremus: in 'Treatise on materials science and technology', (ed. M. Tomozawa and R. H. Doremus), Vol. 17, 41–69; 1979, New York, Academic.
2. B. M. J. Smets and T. P. A. Lommen: *Phys. Chem. Glasses*, 1983, **24**, (1), 35–36.
3. J. Zarzycki: 'Le verre et l'état vitreux', 391; 1982, Paris, Masson.
4. R. W. Douglas and T. M. M. El-Shamy: *J. Am. Ceram. Soc.*, 1967, **50**, 1–8.
5. H. Scholze: 'Glass, nature, structure and properties', 454; 1991, New York, Springer-Verlag.
6. L. L. Hench: *J. Non-Cryst. Solids*, 1975, **19**, 27–39.
7. C. R. Das and R. W. Douglas: *Phys. Chem. Glasses*, 1967, **8**, (5), 178–184.
8. M. Verità: in 'Scienze e materiali del patrimonio culturale: the material of cultural heritage in their environment', (ed. R. A. Lefèvre), Vol. 8, 119–132; 2006, Bari, Edipuglia.
9. R. G. Newton and D. Fuchs: *Glass Technol.*, 1988, **29**, (1), 43–48.
10. L. L. Hench and D. E. Clark: *J. Non-Cryst. Solids*, 1978, **28**, 83–105.
11. G. A. Cox, O. S. Heavens, R. G. Newton and A. M. Pollard: *J. Glass Studies*, 1979, **21**, 54–75.
12. J. O. Isard and A. R. Patel: *Glass Technol.*, 1981, **22**, (6), 247–250.
13. B. C. Bunker: *J. Non-Cryst. Solids*, 1994, **179**, 300–308.
14. H. Römich: *Riv. Stazione Sper. Vetro*, 2000, **6**, 9–14.
15. M. Perez Y Jorba, L. Mazerolles, D. Michel, M. Rommeluere and J. C. Bahezre: Proc. 1er Coll. Programme Franco-Allemand de Recherche pour la Conservation des Monuments Historiques, Karlsruhe, S. Freiherr von Welck Germany, March 1993, 213–219.
16. J. M. Bettembourg: *Verres Réfractaires*, 1976, **30**, (1), 36–42.
17. R.-A. Lefevre, M. Gregoire, M. Derbez and P. Ausset: *Glass Sci. Technol.*, 1998, **71**, 75–80.
18. J. Sterpenich and G. Libourel: *Techne*, 1997, **6**, 70–78.
19. P. Barbay, J. Sterpenich and G. Libourel: Proc. 2nd Coll. Programme Franco-allemand de Recherche pour la Conservation des Monuments Historiques, J.-F. Filtz, Bonn, Germany, December 1996, 61–71.
20. R. G. Newton and S. Davison: 'Conservation in glass', 318; 1989, Oxford, Butterwort and Heinemann.
21. I. Munier, R.-A. Lefevre and R. Losno: *Glass Technol.*, 2002, **43C**, 114–124.
22. M. Schreiner: *Glass Sci. Technol.*, 1988, **61**, (7), 197–204.
23. H. Römich: in 'The conservation of glass and ceramics', (ed. N. Tennent), 57–65; 1999, London, James & James.
24. M. Melcher and M. Schreiner: *Anal. Bioanal. Chem.*, 2004, **379**, 628–639.
25. M. Melcher and M. Schreiner: *J. Non-Cryst. Solids*, 2006, **352**, 368–379.
26. C. M. Jantzen and M. J Plodinec: *J. Non-Cryst. Solids*, 1984, **67**, 207–223.
27. A. Paul: *J. Mater. Sci.*, 1977, **12**, 2246–2268.
28. R. G. Newton and A. Paul: *Glass Technol.*, 1980, **21**, (6), 307–309.